RAND NATIONAL DEFENSE RESEARCH INSTITUTE

ISSUES WITH

Access to Acquisition Data and Information

IN THE DEPARTMENT OF DEFENSE

Considerations for Implementing the
Controlled Unclassified Information Reform Program

Megan McKernan, Jessie Riposo,
Geoffrey McGovern, Douglas Shontz,
Badreddine Ahtchi

Prepared for the Office of the Secretary of Defense

For more information on this publication, visit www.rand.org/t/RR2221

Library of Congress Cataloging-in-Publication Data is available for this publication.
ISBN: 978-0-8330-9979-2

Support RAND
Make a tax-deductible charitable contribution at
www.rand.org/giving/contribute

www.rand.org

Preface

Acquisition data play a critical role in the management of the U.S. Department of Defense's portfolio of weapon systems. Controlled Unclassified Information (CUI) labels are one of the key methods for protecting sensitive information from disclosure, along with appropriate information security. Mandatory U.S. government–wide policies governing handling of unclassified acquisition information exist because of concerns about exploitation by sophisticated adversaries. Executive Order 13556, signed by then–President Barack Obama on November 4, 2010, established a government-wide program for managing CUI, which includes personally identifiable information, proprietary business information, and law enforcement investigation information, among others. As the CUI executive agent, the National Archives and Records Administration is responsible for addressing over 100 ways of characterizing CUI, which it has done in the September 2016 CUI Federal Register. The rules in this register came into effect on November 14, 2016.

The Office of the Secretary of Defense asked the RAND Corporation to take a closer look at the current state of the CUI program as well as how the new CUI rules might affect acquisition data management within the Defense Acquisition Visibility Environment. This report will provide additional context on a topic that will soon require much work and thought to complete successful implementation.

This analysis builds on three earlier studies on *Issues with Access to Acquisition Data and Information in the Department of Defense.* This report should be of interest to government acquisition professionals, oversight organizations, and, especially, the analytic community.

This research was sponsored by the Office of the Secretary of Defense and conducted within the Acquisition and Technology Policy Center of the RAND National Defense Research Institute, a federally funded research and development center sponsored by the Office of the Secretary of Defense, the Joint Staff, the Unified Combatant Commands, the Navy, the Marine Corps, the defense agencies, and the defense Intelligence Community.

For more information on the RAND Acquisition and Technology Policy Center, see www.rand.org/nsrd/about/atp or contact the director (contact information is provided on the webpage).

Contents

Figures and Tables

Summary

Acquisition data are critical to U.S. Department of Defense (DoD) program-related decision-making, execution, analysis, oversight, and—especially—mandatory reporting to Congress. Yet DoD leadership and analysts can have difficulty accessing and managing these data, both as single elements and in aggregate. Mandatory U.S. government–wide policies governing handling of unclassified acquisition information exist because of concerns about exploitation by sophisticated adversaries. But as previous research has demonstrated,[1] these policies are not centralized and are not well-known to information handlers. In addition to the challenges associated with identifying and marking Controlled Unclassified Information (CUI), the circumstances under which a document should no longer be considered CUI are unclear. This lack of standardization has created a tough balancing act for individual agency managers, who must consider both legitimate information requests and security concerns. More importantly, the situation has caused inefficiency and a lack of information for decisionmakers.

This research builds on previous RAND work in this area by investigating and analyzing two interrelated questions:

- How far along is the DoD acquisition community in adopting the latest federal CUI marking reform effort, and what are the potential challenges to that implementation?
- What are the challenges in managing aggregation of DoD acquisition data?

To answer the first question, we examined the state of National Archives and Records Administration (NARA) CUI reform efforts as of early 2017, as well as ongoing DoD efforts. We held discussions with stakeholders in the Office of the Under Secretary of Defense for Acquisition, Technology, and Logistics (OUSD[AT&L]), OUSD for Intelligence, NARA, and the U.S. Department of Justice. The team also conducted an extensive policy review.

To answer the second question, the team worked to define the current landscape in which the Defense Acquisition Visibility Environment—one of three information systems frequently

---

[1]   Jessie Riposo, Megan McKernan, Jeffrey A. Drezner, Geoffrey McGovern, Daniel Tremblay, Jason Kumar, and Jerry M. Sollinger, *Issues with Access to Acquisition Data and Information in the Department of Defense: Policy and Practice*, Santa Monica, Calif.: RAND Corporation, RR-880-OSD, 2015; Megan McKernan, Jessie Riposo, Jeffrey A. Drezner, Geoffrey McGovern, Douglas Shontz, and Clifford A. Grammich, *Issues with Access to Acquisition Data and Information in the Department of Defense: A Closer Look at the Origins and Implementation of Controlled Unclassified Information Labels and Security Policy*, Santa Monica, Calif.: RAND Corporation, RR-1476-OSD, 2016; Megan McKernan, Nancy Young Moore, Kathryn Connor, Mary E. Chenoweth, Jeffrey A. Drezner, James Dryden, Clifford A. Grammich, Judith D. Mele, Walter T. Nelson, Rebeca Orrie, Douglas Shontz, and Anita Szafran, *Issues with Access to Acquisition Data and Information in the Department of Defense: Doing Data Right in Weapon System Acquisition*, Santa Monica, Calif.: RAND Corporation, RR-1534-OSD, 2017.

used for reporting, oversight, and analysis in defense acquisition—must integrate and aggregate acquisition data. We also reviewed related policy and held discussions with CUI experts.

The report is intended to identify some of the potential challenges to implementation for OUSD(AT&L) so that it can prepare and plan for implementation.

CUI Registry Markings Challenging to Implement, but Most Common Acquisition Labels Accounted for in Reform

Executive Order (EO) 13556 establishes a program for CUI management as a response to increasing security concerns since September 11, 2001.[2] This order works in tandem with related DoD polices. NARA is the CUI executive agent and is responsible for addressing over 100 ways of characterizing CUI. The new rule designates all unclassified material that is subject to any special safeguards or special handling requirements as CUI. All safeguards and handling instructions, whether established by law, regulation, or government-wide policy, must now conform to the CUI program.

In general, all federal agencies are bound by the 24 categories of CUI that NARA has established. Table S.1 lists these categories; those in bold are most likely to be relevant to DoD acquisition work.

Table S.2 compares defense acquisition markings commonly used prior to NARA's marking polices with the newer NARA CUI labels. The comparison demonstrates that the legacy

Table S.1
Approved CUI Categories

Agriculture	Legal
Controlled Technical Information	North Atlantic Treaty Organization
Critical Infrastructure	Natural and Cultural Resources
Emergency Management	**Nuclear**
Export Control	Patent
Financial	**Privacy**
Geodetic Product Information	**Procurement and Acquisition**
Immigration	**Proprietary Business Information**
Information Systems Vulnerability Information	SAFETY Act Information
International Agreements	Statistical
Intelligence	Tax
Law Enforcement	Transportation

SOURCE: NARA, "CUI Registry—Categories and Subcategories," webpage, last reviewed November 15, 2017o.
NOTE: Bold entries are relevant to the DoD acquisition community.

---

[2]   EO 13556, *Controlled Unclassified Information*, Washington, D.C.: The White House, 3 C.F.R. 68675, November 4, 2010.

Table S.2
Legacy DoD Labels and New NARA CUI Labels

Legacy DoD Labels	NARA Compliant?	Potential Action Required
Business Sensitive	NO	Can no longer be used; New label could be PROPIN, Procurement and Acquisition (PROCURE), or Proprietary Business Information–Manufacturer (MFC)
Competition Sensitive	NO	Can no longer be used; Could be PROPIN, PROCURE, or MFC
For Official Use Only (FOUO)	NO	Can no longer be used; likely switches to CUI Basic (unless covered by specific regulation)
Pre-Decisional	NO	Can no longer be used as a banner marking or portion marking[a]
Proprietary Information (PROPIN)	ALMOST	Must determine whether CUI Basic or CUI Specified label applies
Source Selection Sensitive	ALMOST	New label is likely CUI–SP//SSEL
Technical Distribution Statement (TDS)	ALMOST	TDSs are now called Limited Dissemination Control Markings (LDCMs) and are required in addition to CUI banner markings for some types of CUI

NOTE: CUI–SP//SSEL = Controlled Unclassified Information—Specified//Procurement and Acquisition-Source Selection
[a] Pre-Decisional can be used as a "Supplemental Administrative Marking" but "may not be used to control CUI and may not be commingled with or incorporated into the CUI Banner Marking or Portion Markings" (NARA, 2016c, p. 21).

labels, though now disallowed upon implementation, are generally subsumed in the new categories and subcategories of CUI that NARA has identified as part of the new regulation.

The review of NARA CUI markings pertinent to the DoD acquisition community suggests that all broadly applicable classes of CUI that the community regularly handles, produces, or acquires are covered. In particular, there are no glaring gaps between what NARA deems CUI and the data that the DoD acquisition community handles.

Implementation of CUI Reform Will Be Significant for DoD

The CUI reform effort was initially meant to increase sharing of materials within the government by standardizing CUI markings and dissemination restrictions. This, in theory, would help government employees know what could be shared, with whom, and when.

However, **recent CUI efforts have taken on the tenor of a security protocol—which may inadvertently exacerbate sharing problems.** This new process has parallels in the classified and information security protocols, including similar banner marking requirements (including an ongoing discussion of portion marking for CUI documents) and enforcement mechanisms for noncompliance to be managed by agency security officers. Potentially, each agency will also have their own office for recording and collecting security incidents and for auditing materials. All of this will require additional security reviews and personnel who can triage materials.

Implementation of the new CUI program is destined to be disruptive. One reason for this is that the new protocols negate and replace what have become customary, perhaps somewhat intractable CUI labeling practices. Notably, the FOUO label, widely used on DoD documentation, is now prohibited. Its replacement sometimes requires a more intensive process for choosing an appropriate label that reflects the information's specific content (a class of CUI known as CUI Specified). Legacy materials also may need to be relabeled, further disrupting DoD practices. Relabeling of legacy information could require contacting the original owners of the information or creating a new adjudication process—a time-intensive effort, to say the least. When compared with the precursor use of labels on CUI material (like FOUO), the new process is much more descriptive and prescriptive.

Given the emphasis on the importance and specificity of labeling information, training is likely to be extensive and include both DoD employees and contractors. Every DoD employee and contractor will need to receive training on the new marking schemas. If DoD decides to implement a portion-marking requirement, training will be more complicated, because not all DoD employees and contractors are familiar with portion marking.

Implementation is currently unfunded, and it is unclear what the financial burden of implementation will be. In particular, this will require potentially significant changes to information systems, depending on the ease of integrating the new set of markings into preexisting materials. This may be particularly difficult, given that not all structured and unstructured data have markings in the current CUI protocol.

Several commonly used labels on acquisition information are no longer permitted, which will leave DoD employees and contractors looking for the next "FOUO." In Chapter Two, we conduct a crosswalk to better understand which labels will be eliminated. In the post-CUI regulation era, "Business Sensitive," "Competition Sensitive," "For Official Use Only," and "Pre-Decisional" are largely noncompliant, as only NARA-approved banner and portion markings can now be used on CUI. In response, DoD will have to redesign its labeling processes to match the CUI registry–approved labels. The crosswalk of the legacy DoD labels with the new NARA labels reveals that legacy labels, although now disallowed, are generally subsumed in the new categories and subcategories of CUI that NARA has identified as part of the new regulation. In other words, the information that the DoD acquisition community regularly handles, produces, and acquires is adequately covered by the CUI registry. The potential difficulty comes from the need to learn the details and definitions of the new categories, identify which are most relevant to the DoD acquisition community, and learn a new labeling regime.

A Still-Unclear Path Forward

In light of the CUI reform effort's shift toward centralization and standardization of the labels for CUI, the biggest concern, ultimately, appears to be that there must now be a process of learning the new categories, subcategories, labels, and LDCMs. We provide an overview of what NARA requires, but many details are still evolving or have yet to be established. Equally important is the cultural and practical deprogramming of the former markings and previously established ways within DoD of handling CUI material.

However, the review of the new CUI Registry markings conducted herein—and, in particular, the crosswalk of the most-common DoD acquisition CUI markings from the past—

indicate that there is a high degree of overlap in the content, if not the nomenclature, of the labels. This suggests that the major thrust of controlling CUI is going to continue at a robust level. The transition period for converting practice (and legacy materials) to the new CUI Registry standard will be uncomfortable, as such changes often are. There are not, however, glaring gaps in what OUSD(AT&L) wishes to control, in what NARA deems CUI, or in who is responsible at the agency level for establishing efficient policy for the future.

Still, there will be trouble in translating policy into practice. Even detailed, clear policy can fail to produce consistent results when it comes to labeling material like CUI. For example, in DoD, FOUO was, by policy, explicitly tied to Freedom of Information Act exemptions. This connection created relatively well-defined boundaries for what was and was not required to be marked. But as shown in previous RAND studies, significant confusion and misperception, particularly about FOUO, exists among DoD personnel, despite the fact that official DoD policies were clear.[3]

Ultimately, federal government agencies should closely monitor the implementation of this effort. It is possible that the best intentions still will not produce the intended benefits of CUI, as this is a massive effort that may be implemented unevenly across the government. In addition, over time, CUI implementation has shifted to a security posture, not a sharing posture.

Security Challenges Related to Aggregated Data Are Perceived as High, but Guidance Exists

DoD holds a tremendous amount of acquisition data. Yet those data are of limited value in the disparate and unstructured forms in which they are often stored. The growing use of so-called "big data" tools exemplifies the potential value of aggregating acquisition data. Further, data aggregation helps create a historical record to draw on for program execution and new program development.

Despite these benefits, the acquisition community, and DoD more generally, are concerned that adversaries could access aggregated information. While acquisition leadership may find that newly aggregated data enable them to present more-robust "state of the acquisition portfolio" reports to Congress, adversaries too may be able to get a fuller understanding of DoD assets. Information managers across DoD expressed additional concerns: (1) the process of examining all potential combinations of information takes significant effort, and (2) the concept of *aggregation* is hard to grasp without clear examples.

Security challenges are recognized in EO 13526, *Classified National Security Information*.[4] The language in this order, while broad in scope, allows the original classification authority to withhold any information that *reasonably could be expected* to damage national security—as long as the potential damage can be identified. This scenario would certainly include specific data elements within a compilation. In addition, the Information Security Oversight Office, DoD, the Department of Commerce, and the Department of Homeland Security have all released policy guidance on classification by compilation. Each of these stresses how infor-

---

[3] Riposo et al., 2015; McKernan et al., 2016; McKernan et al., 2017.

[4] EO 13526, *Classified National Security Information*, Washington, D.C.: The White House, December 29, 2009.

mation may be deidentified (e.g., by removing company names) or summarized (sometimes referred to as "aggregation") for analytic purposes to protect potentially sensitive information.

Options for Moving Forward

As it currently stands, implementation of CUI reform within DoD will have significant effects on the management and handling of acquisition data. Lack of participation could lead to major challenges in OUSD(AT&L)'s day-to-day operations,[5] so OUSD(AT&L) should consider the following options:

- **Identify a point of contact to help advise and transition OUSD(AT&L) to the new marking protocol.** This function could be assigned to someone who already has a role in the organization. For example, this role can be assigned to a senior member of the staff as a collateral duty during the transition.
- **Actively engage in discussions with Undersecretary of Defense for Intelligence (USD[I])** because USD(I) will benefit from understanding OUSD(AT&L)'s needs while revising DoD Manual 5200.1, Vol. IV. OUSD(AT&L)'s participation in reforms to date has been limited mostly to private-sector and technical information.
- **Work closely with NARA as needed** to understand some of the current guidance that has been issued at the federal level.
- **Hold small working groups with the military services and DoD functions** (e.g., Comptroller) to further understand the implications of this effort.
- **Begin to identify training resource requirements.** OUSD(AT&L) may want to create its own focused training for the CUI categories staff are most likely to use, rather than rely solely on the broader DoD implementation training.
- **Wait to implement until USD(I) completes the guidance per USD(I)'s strong recommendation.** Several key pieces of implementation (e.g., portion marking) are still being discussed.
- **Carefully monitor changes** to both the CUI registry and any potential changes to the overall federal CUI strategy by the Trump administration.
- In regard to data aggregation, **Deputy Director, Enterprise Information in OUSD(AT&L) should consider using the National Institute of Standards and Technology's aggregation tool** described in Chapter Three as a mechanism for systematically combing through the information systems that it currently manages for potential sensitivities arising from aggregation.

---

[5] Section 901 of the fiscal year 2017 National Defense Authorization Act eliminates the position of Under Secretary of Defense for Acquisition, Technology, and Logistics and creates two new principal positions: an Under Secretary for Research and Engineering and an Under Secretary for Acquisition and Sustainment (Public Law 114–328, Section 901, National Defense Authorization Act for Fiscal Year 2017, December 23, 2016). This analysis did not formulate conclusions on how these new positions might affect the implementation of the new CUI program.

Acknowledgments

We would like to thank the sponsor of this study, Mark Krzysko, Deputy Director, Acquisition Resources and Analysis, Enterprise Information, within the Office of the Under Secretary of Defense for Acquisition, Technology, and Logistics for his instrumental support and guidance throughout the study. In addition, we would like to thank Joseph Alfano for helping during the initial stages of the study. We also thank Mark Hogenmiller and Kaylee Kehler, who helped facilitate discussions with Controlled Unclassified Information (CUI) experts and provided some additional background information on this topic. We would also like to thank the CUI experts within the U.S. Department of Defense, the National Archives and Records Administration, and the U.S. Department of Justice with whom we held discussions.

We are very grateful to the formal peer reviewers of this document, Stephanie Young and Jordan Bell, who helped improve it through their thorough reviews. We also thank Maria Falvo for her assistance during this effort.

Finally, we would like to thank the director of the RAND Acquisition and Technology Policy Center, Cynthia Cook, and the associate director, Christopher Mouton, for their insightful comments on this research.

Abbreviations

AIR	Acquisition Information Repository
ARA/EI	Acquisition Resources and Analysis, Enterprise Information
BUDG	Financial-Budget
CFR	Code of Federal Regulations
CIO	Chief Information Officer
CRIT	Critical Infrastructure
CTI	Controlled Technical Information
CUI	Controlled Unclassified Information
DAMIR	Defense Acquisition Management Information Retrieval
DAVE	Defense Acquisition Visibility Environment
DCRIT	DoD Critical Infrastructure Security Information
DHS	Department of Homeland Security
DoD	U.S. Department of Defense
DoDI	Department of Defense Instruction
DoDM	Department of Defense Manual
DOJ	U.S. Department of Justice
EI	Enterprise Information
EO	Executive Order
FAR	Federal Acquisition Regulation
FFRDC	federally funded research and development center
FNC	Financial
FOIA	Freedom of Information Act

FOUO	For Official Use Only
ISOO	Information Security Oversight Office
ISVI	Information Systems Vulnerability Information
LDCM	Limited Dissemination Control Marking
MDAP	Major Defense Acquisition Program
MFC	Proprietary Business Information–Manufacturer
NARA	National Archives and Records Administration
NIST	National Institute of Standards and Technology
OCA	original classification authority
OUSD(AT&L)	Office of the Under Secretary of Defense for Acquisition, Technology, and Logistics
PROCURE	Procurement and Acquisition
PROPIN	Proprietary Information
PRVCY	Privacy
SBU	Sensitive but Unclassified
TDS	Technical Distribution Statement
U.S.C.	United States Code
US-CERT	U.S. Computer Emergency Readiness Team
USD(AT&L)	Under Secretary of Defense for Acquisition, Technology, and Logistics
USD(I)	Undersecretary of Defense for Intelligence

CHAPTER ONE

Introduction

U.S. Department of Defense (DoD) acquisition leadership, acquisition professionals, and their supporting analysts should have access to acquisition data to support decisionmaking, execute acquisition programs, and conduct effective analyses. Access to acquisition data also helps support the need for acquisition, budgetary, and cost efficiencies across DoD—a top leadership priority. These data serve as the foundation for Under Secretary of Defense for Acquisition, Technology, and Logistics (USD[AT&L]) insight and decisionmaking on the acquisition portfolio and mandatory reporting to Congress. Acquisition data also enhance oversight and management through the utilization of authoritative data. The accumulation of this large amount of acquisition data provides opportunities for understanding the operation of the Defense Acquisition System through empirical analysis and data analytics.

However, the ability to access and manage acquisition data as both elements and in aggregate has proven problematic. Due to concerns about unauthorized disclosure, aggregation, and exploitation by sophisticated adversaries, for years, government-wide policies have mandated that agencies protect various types of unclassified information. Before 2010, there had not been a single congressional enactment that created a broad class of sensitive but unclassified material. However, in 2010, Executive Order (EO) 13556 established a program for managing Controlled Unclassified Information (CUI).[1] CUI includes personally identifiable information; proprietary business information; and law enforcement investigations, among others. This program emphasizes security—as have past laws, regulations, and policies—but also stresses the importance of government-wide sharing through standardization of labeling. As the CUI executive agent, the National Archives and Records Administration (NARA) is responsible for addressing over 100 ways of characterizing CUI. EO 13556 established the CUI program to help simplify the handling of unclassified information:

> [The CUI program] is a system that standardizes and simplifies the way the Executive branch handles unclassified information that requires safeguarding or dissemination controls, pursuant to and consistent with applicable law, regulations, and government-wide policies. The program emphasizes the openness and uniformity of government-wide practices. Its purpose is to address the current inefficient and confusing patchwork that leads to inconsistent marking and safeguarding as well as restrictive dissemination policies, which are often hidden from public view.

---

[1] EO 13556, *Controlled Unclassified Information*, Washington, D.C.: The White House, 3 C.F.R. 68675, November 4, 2010.

The President has designated [NARA] as the CUI Executive Agent (EA). In this role, NARA has the authority and responsibility to oversee and manage the implementation of the CUI program and will issue policy directives and publish reports on the status of agency implementation.[2]

DoD has not yet implemented the federal CUI policy. On one hand, the lack of implementation may pose a risk to properly securing business sensitive and other types of sensitive unclassified information. On the other hand, imposing federal CUI standards that do not adequately reflect DoD's distinctive processes and requirements may create significant inefficiencies in operations.

Within DoD, there is no common definition and only limited standardized protocols describing the circumstances under which information should be marked CUI, the criteria for identifying information as CUI, and the circumstances under which information should no longer be considered CUI. One result is that security-oriented data managers are incentivized to restrict access to such information. This tendency to restrict access to CUI for security reasons imposes barriers to sharing data for the legitimate and beneficial purposes described above.

In our earlier work on managing and sharing acquisition data,[3] we found a complex set of rules and practices governing CUI labels and security policies for acquisition data, as illustrated in Figure 1.1. The Deputy Director, Acquisition Resources and Analysis, Enterprise Information (ARA/EI), within the Office of the Under Secretary of Defense for Acquisition,

Figure 1.1
Influences on Access to Acquisition Data

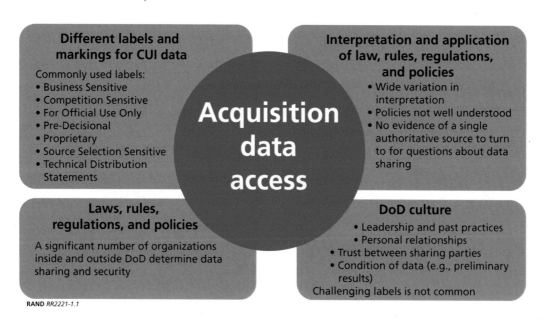

RAND RR2221-1.1

---

[2]   NARA, Controlled Unclassified Information Office, *What Is CUI? Answers to the Most Frequently Asked Questions*, Washington, D.C., 2011.

[3]   Jessie Riposo, Megan McKernan, Jeffrey A. Drezner, Geoffrey McGovern, Daniel Tremblay, Jason Kumar, and Jerry M. Sollinger, *Issues with Access to Acquisition Data and Information in the Department of Defense: Policy and Practice*, Santa Monica, Calif.: RAND Corporation, RR-880-OSD, 2015.

Technology, and Logistics (OUSD[AT&L]), is leading efforts to manage these data as part of its core mission to improve acquisition data collection and management. RAND has been supporting these efforts since 2013 through the exploration of challenges in managing and accessing acquisition data. This study builds on knowledge from three prior studies (see Figure 1.2).

In Phase 1 of our studies, we found that the management and sharing of acquisition data are subject to the interaction and interpretation of a large number of laws, regulations, and policies; CUI labels; and DoD culture, among other influences. This complex environment leads to a host of inefficiencies for those who manage and utilize these data, resulting in some acquisition professionals not getting data they need for their assigned duties or not getting data and information in an efficient manner.[4]

Furthermore, balancing security and transparency has been an ongoing challenge. In Phase 2, we took a closer look at several key sources of inefficiency by evaluating how marking and labeling CUI procedures, practices, and security policies affect access to and management of acquisition oversight data.[5] This builds on Phase 1 by examining in more detail issues with sharing proprietary information, using CUI labels, and implementing security policy in the Acquisition Information Repository (AIR) and the Defense Acquisition Management Information Retrieval (DAMIR) systems.

After examining policies, procedures, and practices and associated problems, we were then asked to help identify how available data can help assist defense acquisition decisionmaking.[6] In

Figure 1.2
Prior RAND Studies Investigating Acquisition Data Challenges

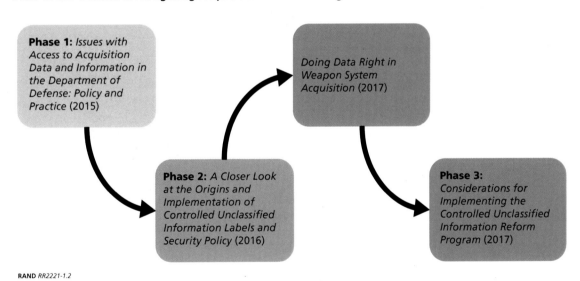

RAND RR2221-1.2

4 Riposo et al., 2015.

5 Megan McKernan, Jessie Riposo, Jeffrey A. Drezner, Geoffrey McGovern, Douglas Shontz, and Clifford A. Grammich, *Issues with Access to Acquisition Data and Information in the Department of Defense: A Closer Look at the Origins and Implementation of Controlled Unclassified Information Labels and Security Policy,* Santa Monica, Calif.: RAND Corporation, RR-1476-OSD, 2016.

6 Megan McKernan, Nancy Young Moore, Kathryn Connor, Mary E. Chenoweth, Jeffrey A. Drezner, James Dryden, Clifford A. Grammich, Judith D. Mele, Walter T. Nelson, Rebeca Orrie, Douglas Shontz, and Anita Szafran, *Issues with*

particular, we documented where some data reside, who can access the data, and who owns the information in 21 information systems.

Approach

Our work for Phase 3 of this research on managing and handling DoD acquisition data included policy analysis, structured discussions with government personnel, and a literature review to further understand and evaluate the NARA CUI reform effort's potential implementation consequences. We executed our work through two main tasks.

Task 1: Define the current state of CUI. To understand the current state of CUI reforms, we reviewed the proposed NARA CUI reforms and ongoing DoD efforts through discussions with appropriate stakeholders in OUSD(AT&L), OUSD(Intelligence), NARA, and the U.S. Department of Justice (DOJ) as well as conducting a policy review. In addition, we analyzed whether proposed NARA CUI reforms meet the needs of DoD acquisition and then evaluated the proposed labeling processes and procedures through a crosswalk between legacy and new commonly used acquisition CUI labels.

Task 2: Determine how aggregation or compilation of CUI is defined in current laws, regulation, and policy for acquisition data. Given the role of Enterprise Information (EI) as an integrator and manager of acquisition information, we reviewed current policy involving aggregation of CUI along with the benefits of and security concerns about aggregation. We also discussed current aggregation practices involving this type of information through discussions with CUI experts.

Organization of This Report

Chapter Two describes the current CUI reform effort. We include both background and analysis on how this effort will fit into the current acquisition data environment. Chapter Three provides a policy review and analysis of the aggregation of CUI. Chapter Four provides our conclusions and options for OUSD(AT&L) to consider from this analysis. Finally, Appendix A provides additional information on the nine NARA CUI categories that represent the topical areas where acquisition work could touch upon CUI materials.

Access to Acquisition Data and Information in the Department of Defense: Doing Data Right in Weapon System Acquisition, Santa Monica, Calif.: RAND Corporation, RR-1534-OSD, 2017.

CHAPTER TWO

Overview and Analysis of the Current CUI Reform Effort

This chapter provides an overview of the history and current status of the CUI reform effort. We also provide a crosswalk of how the new CUI labeling scheme may work with or replace some commonly used acquisition labels.[1] Finally, we discuss the path forward as it pertains to acquisition data management within DoD.

The CUI effort is intended to standardize labels and control markings across the government, which in theory should benefit those who need to share information. The proposed CUI scheme somewhat parallels the approach for classified information, as some categories of information require specific handling, and there is an ongoing discussion about requiring portion marking. The largest effects on daily practice in DoD are likely:

1. the shift away from the ubiquitous "For Official Use Only" (FOUO) label
2. the challenges of marking new information
3. the handling and relabeling (or delabeling) of a vast amount of legacy information.

CUI markings do not affect determinations of whether information can be released to the public. Public disclosure remains a decision based on a detailed review of the content of the document, independent of any label.

Readers should note that the content of this chapter was current as of July 2017. DoD is undergoing implementation review of the CUI program and devising ways to tailor the requirements to the DoD environment, and subsequent developments by various DoD offices may revise programmatic details in the future.

History of the CUI Program

While concerns about unclassified information sharing, notably Sensitive but Unclassified (SBU) information, have existed since at least the Cold War, the roots of the new CUI policy can be traced to September 11, 2001.[2] Some of the roots of SBU being tied to security can be found in National Security Decision Directive 145:

> In 1984, National Security Decision Directive 145 (NSDD-145) directed that "sensitive, but unclassified, government or government-derived information, the loss of which

---

[1] McKernan et al., 2016.

[2] OMB Watch, *Controlled Unclassified Information: Recommendations for Information Control Reform*, Washington, D.C.: Center for Effective Government, July 2009, p. 3.

could adversely affect the national security interest" should be "protected in proportion to the threat of exploitation and the associated potential damage to the national security." NSDD-145 did not define the term, "sensitive, but unclassified," but explained that even unclassified information in the aggregate can "reveal highly classified and other sensitive information . . ." harmful to the national security interest.[3]

The 9/11 attacks led to the creation of the National Commission on Terrorist Attacks upon the United States, otherwise known as the 9/11 Commission, to comprehensively investigate the circumstances surrounding the events. One of the key findings within the commission's final report was the following:

> Information was not shared, sometimes inadvertently or because of legal misunderstandings. Analysis was not pooled. Effective operations were not launched. Often the handoffs of information were lost across the divide separating the foreign and domestic agencies of the government.[4]

At the time, there was no single entity responsible for government-wide concepts and information-sharing standards. The government-wide culture of compartmentalization and overclassification of information was publicly scrutinized after 9/11 and led to the change in policy. The report recommended that the executive branch lead the effort to resolve the legal, policy, and technical issues across agencies to create a trusted information network and to bring the national security institutions into the information revolution.[5]

One year after the 9/11 Commission report was released, then–President George W. Bush sent out a memorandum to the heads of executive departments and agencies regarding support for the new information-sharing environment. The document established the Information Sharing Policy Coordination Committee for interagency action and directed the standardization of procedures for SBU information throughout the government.[6] This memorandum would ultimately set the stage for the creation of CUI. Figure 2.1 shows a timeline of the key events in the CUI reform effort.

In May 2008, Bush signed a memorandum for the designation and sharing of CUI. The policy document implemented the term *controlled unclassified information* as the single label for all SBU information throughout the U.S. federal government information enterprise. The memorandum defines CUI as

> unclassified information that does not meet the standards for National Security Classification under Executive Order 12958, as amended, but is (i) pertinent to the national interests of the United States or to the important interests of entities outside the Federal Govern-

---

[3]   Genevieve J. Knezo, *CRS Report for Congress: "Sensitive But Unclassified" and Other Federal Security Controls on Scientific and Technical Information: History and Current Controversy*, Washington, D.C.: Congressional Research Service, February 20, 2004, p. CRS-11.

[4]   National Commission on Terrorist Attacks upon the United States, *The 9/11 Commission Report: Final Report of the National Commission on Terrorist Attacks upon the United States*, Washington, D.C.: U.S. Government Printing Office, July 22, 2004, p. 353.

[5]   National Commission on Terrorist Attacks upon the United States, 2004, p. 418.

[6]   George W. Bush, "Guidelines and Requirements in Support of the Information Sharing Environment," memorandum for the Heads of Executive Departments and Agencies, Washington, D.C., December 16, 2005.

Figure 2.1
Key Events in the CUI Reform Effort

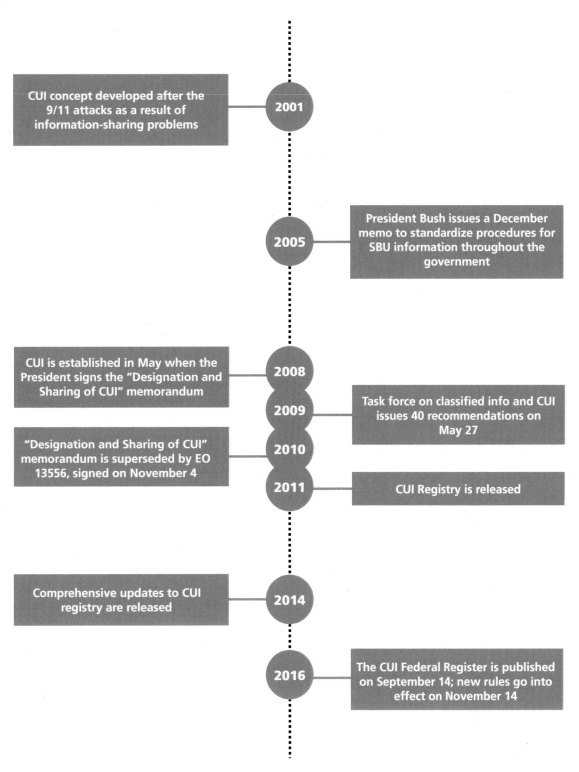

SOURCE: NARA, "Chronology," webpage, last updated December 14, 2016d, and discussions with CUI subject-matter experts.

RAND *RR2221-2.1*

ment, and (ii) under law or policy requires protection from unauthorized disclosure, special handling safeguards, or prescribed limits on exchange or dissemination.[7]

The document designated NARA as the executive agent of this order and established an initial framework for this new classification. In response, the Archivist of the United States issued a memorandum shortly after establishing the CUI office within NARA that would be responsible for developing and issuing CUI policy standards.[8]

Given that the establishment of CUI arrived in the latter months of Bush's tenure, this initiative was something that then–President Barack Obama needed to address in his first term. Accordingly, Obama ordered the Attorney General and the Secretary of Homeland Security to lead an interagency task force on classified information and CUI, which reevaluated the status quo of the information-sharing environment and directed next steps for CUI.[9] The subsequent task force report provided a host of recommendations, which included expanding the scope of the CUI framework to all SBU information, not just terrorism-related information.[10]

EO 13556 Created the New CUI Program

These recommendations laid the foundation for Obama's EO 13556 on CUI. Issued in November 2010, it established an

> open and uniform program for managing information that requires safeguarding or dissemination controls pursuant to and consistent with law, regulations, and Government-wide policies, excluding information that is classified under . . . the Atomic Energy Act . . . as amended.[11]

The order mandated that each agency submit to NARA within 180 days its proposed categories and subcategories of CUI, and that NARA create a public CUI registry within 365 days that records all authorized CUI categories, subcategories, associated markings, and applicable procedures.[12] DoD participated in the process of creating the CUI Registry, with NARA in the lead, though interviews with DoD staff revealed that this process was driven largely by NARA. DoD expressed concerns during the formulation process, including the difficulties of handling portion-marking requirements, the inclusion of labels for email communications, and the relabeling of legacy materials.

The CUI registry was released in 2011, with initial categories and subcategories collected from the various government agencies. A further-refined version of the registry was released in

---

[7]   George W. Bush, "Designation and Sharing of Controlled Unclassified Information (CUI)," memorandum for the Heads of Executive Departments and Agencies, Washington, D.C., May 7, 2008.

[8]   Allen Weinstein, "Establishment of the Controlled Unclassified Information Office," memorandum for the Heads of Executive Departments and Agencies, Washington, D.C., May 9, 2008.

[9]   Barack Obama, "Classified Information and Controlled Unclassified Information," memorandum for the Heads of Executive Departments and Agencies, Washington, D.C., May 27, 2009b.

[10]   U.S. Task Force on Controlled Unclassified Information, *Report and Recommendations of the Presidential Task Force on Controlled Unclassified Information*, Washington, D.C., August 25, 2009, p. 10.

[11]   EO 13556, 2010.

[12]   EO 13556, 2010.

August 2014, which chiefly included additional category information to identify authorities with specific handling requirements based on federal regulation or policy.[13]

Most recently, in September 2016, the Information Security Oversight Office (ISOO) published Code of Federal Regulations (CFR), Title 32, Section 2002, Controlled Unclassified Information (final rule effective as of November 14, 2016).

The following section of this report will expand on the implications of this new rule and the current status of the CUI program.

Status of the CUI Program

As reviewed above, the government has long been interested in the careful handling and control of unclassified material. Data (including documents and electronic forms) can contain sensitive information that warrant special processes for dissemination to the public, for sharing with other government agencies, and even for distribution within agencies. These concerns exist even though the documents and data do not fall within the classified realm governed by the 5200 series of DoD directives, or more broadly, under EO 13526.[14]

The current result of the revision of national policy regarding CUI is a centrally developed system of grouping and labeling CUI that is administratively run by NARA. Fulfilling the direction given by the president in EO 13556, NARA organized a process of harmonizing CUI categories and, through the ISOO, published regulation 32 CFR 2002 (the final rule effective as of November 14, 2016). The rule applies to all executive branch agencies, as well as all other organizations that "handle, possess, use, share, or receive CUI—or which operate, use, or have access to Federal information and information systems on behalf of an agency." In substance, the rule comprehensively "establish[es] policy for agencies on designating, safeguarding, disseminating, marking, decontrolling, and disposing of CUI, self-inspection and oversight requirements, and other facets of the Program."[15]

The new rule designates all unclassified material that is subject to any special safeguards or special handling requirements as CUI. It is the safeguarding and dissemination procedures that determine whether information is CUI. All safeguards and handling instructions, whether established by law, regulation, or government-wide policy, must now conform to the CUI program that NARA has established. According to 32 CFR 2002.4(h):

> Law, regulation, or Government-wide policy may require or permit safeguarding or dissemination controls in three ways: Requiring or permitting agencies to control or protect the information *but providing no specific controls, which makes the information CUI Basic*; requiring or permitting agencies to control or protect the information and providing *specific controls for doing so, which makes the information CUI Specified*; or requiring or permitting agencies to control the information and specifying only some of those controls, *which makes the information CUI Specified, but with CUI Basic controls where the authority does not specify* (emphasis added).[16]

---

[13] NARA, "CUI Registry—Categories and Subcategories," webpage, last reviewed November 15, 2017o.

[14] EO 13526, *Classified National Security Information*, Washington, D.C.: The White House, December 29, 2009.

[15] Code of Federal Regulations, Title 32, Section 2002, Controlled Unclassified Information, September 14, 2016.

[16] Code of Federal Regulations, Title 32, Section 2002.4(h), Definitions, September 14, 2016.

The regulation develops the first major dimension of the CUI program NARA established: namely, the distinction between two different types of CUI: *CUI Basic* and *CUI Specified*. This distinction bears further explanation because it is the crucial determinant of the proper labels to be applied to the CUI.

CUI Basic and CUI Specified: Handling Instructions and Labels

The distinction between CUI Basic and CUI Specified is important for two reasons. First, the difference reflects whether or not there are specific handling procedures for the information, as established in law, regulation, or government-wide policy. Both types are internal administrative handling controls, as established in law or regulation, but only CUI Specified has handling or dissemination instructions (and sanctions for misuse) additionally established in law or policy.

According to NARA, "[w]hether CUI is Basic or Specified is determined by the applicable Safeguarding and/or Dissemination Authority for that CUI."[17] For example, contractor bid or proposal and source selection information is identified as CUI Specified because 48 CFR 3.104-4 creates specific instructions on who can disclose, handle, access, and otherwise use source selection information.[18] The key to the distinction is the presence or absence of specific handling instructions and/or sanctions. As discussed below, however, the application of these distinctions when labeling information may prove difficult if information handlers are not well-versed in the safeguarding and/or dissemination authorities (which can be rather technical).

The second important distinction between CUI Basic and CUI Specified concerns the labeling regime NARA requires. All CUI must be labeled as such; the nature and format of the label will differ in accordance with the content of the CUI. Detailed instruction on how to mark CUI is published in NARA's *Marking Controlled Unclassified Information*.[19] At the core of the labeling scheme are banner markings that identify the material as CUI. For CUI Basic, ISOO instructs that the material can be banner marked as "CONTROLLED," "CUI," or "CUI BASIC."[20] Those labels are appropriate for material that by law, regulation, or government-wide policy has some level of internal administrative control applied to it.

In the case of CUI Specified information, further markings for category and subcategory are required. These categories and subcategories refer to types of content that NARA has identified, approved, and listed in the CUI Registry (an online source that contains information on the CUI program).[21] There are currently 24 categories and 89 subcategories of CUI (see Table 2.1). All material that is CUI Specified must reference the category and/or subcategory in which its content falls. Agencies may, but are not required to, promulgate instructions that CUI Basic material also must bear markings that identify the category and/or subcategory of

---

[17]   NARA, "CUI Registry: Procurement and Acquisition," webpage, last reviewed July 27, 2017k.

[18]   NARA, 2017k.

[19]   NARA, 2016c.

[20]   Note that we do not detail every aspect of the ISOO marking guidance in this report. Here, we provide a top-level review of the most important details of the new CUI program. Full details of the labeling protocol can be found in NARA, 2016c, pp. 7 and 9.

[21]   NARA, "CUI Registry: Critical Infrastructure-DoD Critical Infrastructure Security Information," webpage, last reviewed June 19, 2017a; NARA, 2017o.

Table 2.1
The 24 Approved CUI Categories

Agriculture	Legal
Controlled Technical Information	North Atlantic Treaty Organization
Critical Infrastructure	Natural and Cultural Resources
Emergency Management	**Nuclear**
Export Control	Patent
Financial	**Privacy**
Geodetic Product Information	**Procurement and Acquisition**
Immigration	**Proprietary Business Information**
Information Systems Vulnerability Information	SAFETY Act Information
International Agreements	Statistical
Intelligence	Tax
Law Enforcement	Transportation

SOURCE: NARA, "CUI Registry—Categories and Subcategories," webpage, last reviewed November 15, 2017o.

the content. The categories in bold in the table are most likely to be used by those in DoD acquisition.

These 24 categories were created by NARA in consultation with the government agencies. They represent the broad areas that may contain CUI. Of course, not every piece of information or document relating to these categories is CUI: agencies are asserting the need for internal administrative controls on some information because of perceived sensitivity. Descriptions for each category are provided by NARA at the CUI Registry and range in specificity. For example, the description for the CUI agriculture category is fairly general: "information related to the agricultural operation, farming or conservation practices, or the actual land of an agricultural producer or landowner." On the other hand, the description for Controlled Technical Information is in-depth:

> Controlled Technical Information means technical information with military or space application that is subject to controls on the access, use, reproduction, modification, performance, display, release, disclosure, or dissemination. Controlled technical information is to be marked with one of the distribution statements B through F, in accordance with Department of Defense Instruction 5230.24, "Distribution Statements of Technical Documents." The term does not include information that is lawfully publicly available without restrictions. "Technical Information" means technical data or computer software, as those terms are defined in Defense Federal Acquisition Regulation Supplement clause 252.227-7013, "Rights in Technical Data—Noncommercial Items" (48 CFR 252.227-7013). Examples of technical information include research and engineering data, engineering drawings, and associated lists, specifications, standards, process sheets, manuals, technical reports, technical orders, catalog-item identifications, data sets, studies and analyses and related information, and computer software executable code and source code.[22]

---

[22] NARA, 2017o.

Agencies that work with information that falls within approved CUI categories should take care to understand the types of information they possess and to identify whether the information is subject to legal or regulatory controls regarding handling, dissemination, or pertinent sanctions for mishandling the information.[23]

For an example of a banner marking for CUI Specified information, consider the following hypothetical scenario. An OUSD(AT&L) employee has a document that contains proprietary business information relating to a trade secret that a government contractor wanted to keep protected and which the government is required to keep secure per 15 U.S.C. 2055. This document must be considered CUI because it relates to a CUI category; further, it falls into a CUI Specified category due to legal restrictions on government handling of proprietary information.[24] In this example, the document should be labeled at least with a banner marking as follows: CUI//SP-PROPIN.

Such a label indicates that the information is CUI, that the information falls within a specific set of laws regarding the proper handling of the information (SP), and that the content of that information falls in the category of proprietary information (PROPIN). The PROPIN label is mandated by the CUI registry, which has similar abbreviations for each of the categories and subcategories.

NARA and ISOO also have made provisions for including dissemination statements as part of the CUI labeling regime. As with the list of CUI Specified markings (like PROPIN), the CUI Registry promulgates the official list of Limited Dissemination Control Markings (LDCMs) that signal the allowable dispersion of the information. These include no foreign dissemination (NOFORN), federal employees only (FED ONLY), and no dissemination to contractors (NOCON).[25]

Finally, it is worth noting here that while banner marking is required for all CUI materials, portion marking (the marking of each paragraph, figure, table, etc.) is left to the discretion of the implementing agencies at this time. This has provided the benefit of flexibility, but the vice of variety (which is anathema to the original purpose of the CUI registry).

NARA Categories of Importance to the DoD Acquisition Community

With the centralization of the CUI program to NARA, all agencies must comply with a new labeling regime and a single repository for CUI categories. In theory, all agencies are bound by the 24 categories of CUI. However, only certain categories are going to be pertinent to the DoD acquisition community. This section is designed to winnow down those 24 categories and identify those that are germane to professionals within the DoD acquisition community.

Within Table 2.1, all of the 24 categories in the NARA CUI Registry are presented. We highlighted those categories in bold that may be relevant to defense acquisition professionals.

The nine bold categories represent the topical areas where acquisition work could touch upon CUI materials. Because the key effort for information managers will be to understand which information is CUI, Appendix A provides a review of each category, the definition of

---

[23] A discussion of most DoD acquisition community relevant categories and subcategories is provided in the appendix.

[24] "SP" (or "Specified") indicates that the information falls within a specific set of laws regarding the proper handling of the information.

[25] For the full list of the LDCMs, see NARA, "CUI Registry: Limited Dissemination Controls," webpage, last reviewed June 19, 2017b.

the information under its purview, the labels NARA requires, and where necessary, areas of ambiguity.

Security Measures and Safeguarding of CUI

The labeling program included as part of the new CUI regulation, described above, loosely tracks some of the aspects of the classified information security protocols. These similarities include, but are not limited to

- banner marking
- portion marking (required for classified material; optional for CUI)
- the specification of types of material contained in the document or data set
- LDCMs
- designation indicators that identify the agency that has determined the CUI status of the information
- CUI coversheets
- room, area, and container markings that indicate the presence of CUI.

When compared with the precursor use of labels on CUI material (like FOUO), the new process is much more descriptive and prescriptive.

Furthermore, the new NARA regulation and guidance takes a more directed approach to the safeguarding of CUI material; the regulations specifically call upon agencies to establish processes and criteria for investigating the misuse and mishandling of CUI materials. Those agencies are also authorized to take administrative action against agency personnel who do not comply with the CUI protocols (presumably both through failure to comply as well as intentional misuse or disclosure of CUI material). Again, this brings the CUI program closer to an information security protocol than to a sharing regime.

Implementing the CUI Program

Implementation of the new CUI program, while organized by ISOO at NARA, is further delegated to the executive agencies. For DoD, the authority has been given to the Undersecretary of Defense for Intelligence (USD[I]) as the senior agency official in charge of the CUI program—a strong indicator of the parallels with the classified information labeling scheme. USD(I) has been working with ISOO throughout the development of the CUI program and continues to develop new policy for DoD (an updated version of the 5200 series of policy is forthcoming and will incorporate instructions for deployment of the CUI program at DoD). USD(I) established a working group to coordinate efforts to create new DoD policy in accordance with the NARA guidelines and the new regulation. Supporting USD(I) is the DoD Chief Information Officer (CIO), who has responsibility for implementation of the CUI program via digital tools and contractor information systems (per DoD Instruction [DoDI] 8582.01).[26] Under the aegis of USD(I), other DoD offices have been coordinating policy recommendations for how to align the NARA effort with existing and new practices regarding CUI handling.

---

[26] DoDI 8582.01, *Security of Unclassified DoD Information on Non-DoD Information Systems*, Washington, D.C., June 6, 2012.

Assessment of CUI Labeling

In previous RAND research, we examined OUSD(AT&L) offices' handling of CUI material *before* the publication of the final CUI regulation in November 2016. These previous studies aimed to identify which labels were most commonly used to mark CUI and to investigate the basis of those markings in law, regulation, and government-wide policy.[27] Our review concluded that there were about a dozen commonly used labels, but that some of these were not based in law, regulation, or government-wide policy; instead, some labels were customary (such as "business sensitive") and, while related to actual labels specified in law, were nonetheless inventions arising from practice, not policy. Additionally, we found that there was a disconnect between some labels and their purposes. For example, the near-ubiquitous FOUO label was technically a label arising from policy surrounding DoD implementation of the Freedom of Information Act (FOIA). However, in practice, FOUO was (and remains) over-used without reference to the special exemptions to FOIA requests that necessitated the label in the first place. In other words, FOUO was designed as a technical distinction to fulfill a legal need, but the practice is to apply FOUO generously, without regard to the limited legal justification.

The efforts of the previous studies complement the NARA work in many ways and helped to prepare for the ultimate shift to the new CUI regulation. Our efforts to identify whether there was a basis in law, regulation, or government-wide policy for the frequently used labels helped to reveal some mismarking practices that could be useful case studies as DoD begins to implement a new NARA-compliant CUI program. Furthermore, the framework of focusing on law, regulation, and government-wide policy as the basis of labels fits very well with the new requirement that CUI Specified materials have to be linked to the governance regime (because specific handling instruction in law and regulation, including dissemination controls and sanctions for mishandling, trigger the requirement for labeling CUI Specified materials under a stricter and more burdensome rubric).

Nonetheless, there remains a question of how the previously used labels for defense acquisition fit with the new categorical approach required in the CUI regulation. In other words, are the new labels adequate to cover all of the previously used labeling practices, or are there types of information that still fall outside the CUI categories? Relatedly, which labeling practices will no longer be permitted or will need to be revised to be NARA compliant?

Crosswalk of Legacy Labels with NARA CUI Program

To answer these questions, we "crosswalked" the labels identified in our previous work with the new labels required by NARA. We wanted to know which old labels are permitted, reformatted, or restricted from use under the new guidelines. Table 2.2 displays the commonly used labels we addressed in previous research. To be clear, these are the labels that were previously in use (and likely to be found frequently on legacy information), but they are not compliant with the NARA CUI program.

These labels, while not exhaustive, were identified in interviews with OUSD(AT&L) staff as the most common, most important, and most relevant labels for acquisition purposes (excluding all classified material). The table identifies the label; the DoD policy owner, if applicable, who can promulgate regulation of such material; whether the label had a specific basis in law, regulation, or government-wide policy; whether the label was clearly defined in terms

---

[27] Riposo, et al., 2015; McKernan, et al., 2016.

Table 2.2
Common Data Labels, Authorization Basis, and Access Details

Label Placed on Information or Data	DoD Policy Owner	Basis	Defined?	Clear Handling Procedures?	Is Nongovernment Access Allowed?
Business Sensitive	ASD(NII)/ DoD CIO	DoDI 8520.03	Yes	No	Unclear
Competition Sensitive	Undefined	Sample NDA created by OUSD(AT&L)	Yes	Yes	FFRDC; contractor access possible
FOUO	USD(I)	DoDM 5200.01, Vol. 4	Yes, as exemption to FOIA	Yes	FFRDC; contractor access possible
Pre-Decisional	Undefined	FOIA court cases	Yes	No	Unclear
PROPIN	Undefined	FOIA court cases, law, regulation, policy	Yes, for technical data; No, for nontechnical data	Yes, for technical data; No, for nontechnical data	FFRDC; contractor access possible with NDA for technical data; unclear for nontechnical data
Source Selection Sensitive	USD(AT&L)	41 U.S.C. 2102, FAR 2.101, DoD policies	Yes	Yes, "Source Selection Procedures," 2011	FFRDC; contractor access possible
TDS	USD(AT&L); USD(I)	DoDI 5230.24; DoDM 5200.01, Vol. 4	N/A	N/A	FFRDC; contractor access possible

SOURCE: McKernan et al., 2016, p. 20.
NOTES: ASD(NII) = Assistant Secretary of Defense for Networks and Information Integration; DoDM = DoD Manual; FAR = Federal Acquisition Regulation; FFRDC = federally funded research and development center; NDA = nondisclosure agreement; TDS = Technical Distribution Statement.

of the content that was appropriate for such a label; whether there were clear handling instructions for the material bearing such a label; and whether information bearing a label was able to be shared beyond government employees (namely, with government contractors or FFRDCs).

What emerged from this initial cut was a series of labels that some DoD staff were using in an attempt to comply with a nascent internal CUI regime. Information that was intended only for government employees was frequently labeled FOUO. Often, if the material contained business information that was provided through a contractor or manufacturer, the information was labeled Business Sensitive. If the information was background information used to inform a decision to be made at some point in the future, it was often labeled Pre-Decisional (and FOUO). There was little to no oversight of the marking process. Recipients of the information were not told which parts of the information were controlled and had little recourse to challenge labels they thought were inappropriate. In such an environment, the profusion of labels, like Pre-Decisional and Competition Sensitive, which have no basis in DoD policy, is best understood as an attempt by DoD employees to create CUI controls for information that they believed should be controlled.

In the post-CUI regulation era, however, these labels are largely noncompliant; only NARA-approved markings can now be used on CUI. In response, DoD will have to redesign its labeling processes to match the CUI registry–approved labels. To see how closely the legacy DoD labels and the NARA labels coincide, we attempted to match the legacy labels

Table 2.3
Crosswalk of Legacy DoD Labels and New NARA CUI Labels

Legacy DoD Label	NARA Compliant?	Potential Action Required
Business Sensitive	NO	Can no longer be used; New label could be PROPIN, Procurement and Acquisition (PROCURE), or Proprietary Business Information-Manufacturer (MFC)
Competition Sensitive	NO	Can no longer be used; Could be PROPIN, PROCURE, or Proprietary MFC
FOUO	NO	Can no longer be used; likely switches to CUI Basic (unless covered by specific regulation)
Pre-Decisional	NO	Can no longer be used as a banner marking or portion marking[a]
PROPIN	ALMOST	Must determine whether CUI Basic or CUI Specified label applies
Source Selection Sensitive	ALMOST	New Label is likely CUI–SP//SSEL
TDS	ALMOST	TDSs are now called LDCMs and are required in addition to CUI banner markings for some types of CUI

NOTE: CUI–SP//SSEL = Controlled Unclassified Information—Specified//Procurement and Acquisition–Source Selection
[a] Pre-Decisional can be used as a "Supplemental Administrative Marking" but "may not be used to control CUI and may not be commingled with or incorporated into the CUI Banner Marking or Portion Markings" (NARA, 2016c, p. 21).

with a new, required label from NARA's Registry. Table 2.3 first shows whether the old label is NARA compliant and then describes some potential actions that might need to be taken to bring the CUI into alignment with the new regulation.

For example, DoD used to label documents with Business Sensitive, but this is no longer a recognized category. A legacy document with such a label could contain CUI that requires protection, but DoD employees will have to be more specific in their labeling practices going forward. This could include, when appropriate, the use of CUI Basic (the most general label), or the CUI Specified labels PROPIN, PROCURE, or MFC. The new regulation requires a more tailored labeling determination that not only conforms to NARA guidelines, but also provides more information about the nature of the content.

Another useful case to walk through is the use of the FOUO label. As a policy matter, FOUO was a dissemination control label. DoDM 5200.01, Vol. 4, governed the use of the FOUO label and directed the label to be applied to CUI "when the disclosure of the information would reasonably be expected to cause a foreseeable harm to an interest protected by one or more of FOIA Exemptions 2 through 9."[28] FOUO was created as a by-product of the exemptions in FOIA. An edited list of the applicable exemptions, per DoDM 5200.01, Vol. 4, follows:

- **Exemption 2:** Information that pertains solely to the internal rules and practices of the agency that, if released, would allow circumvention of an agency rule, policy, or statute, thereby impeding the agency in the conduct of its mission.

---

[28] DoDM 5200.01, CUI, Vol. 4, Washington, D.C., May 12, 2012.

- **Exemption 3**: Information that is specifically exempted by a statute establishing particular criteria for withholding. The language of the statute must clearly state that the information will not be disclosed.
- **Exemption 4:** Information, such as trade secrets and commercial or financial information obtained from a company on a privileged or confidential basis, that, if released, would result in competitive harm to the company, impair the government's ability to obtain similar information in the future, or impair the government's interest in compliance with program effectiveness.
- **Exemption 5:** Inter- or intra-agency memorandums or letters that contains information considered privileged in civil litigation.
- **Exemption 6:** Information, that, if released, would reasonably be expected to constitute a clearly unwarranted invasion of the personal privacy of individuals.
- **Exemption 7:** Records or information compiled for law enforcement purposes.
- **Exemption 8:** Certain records of agencies responsible for supervision of financial institutions.
- **Exemption 9:** Geological and geophysical information (including maps) concerning wells.[29]

Only for these reasons would FOUO have been an appropriate label. In practice, however, it is unlikely that DoD information was labeled with full understanding of the FOIA exemptions as the trigger for FOUO applicability.

In response to the NARA changes, using an FOUO label is no longer acceptable practice. Information labelers must now be much more cognizant of the content in their possession. What, then, replaces FOUO? The answer is that the appropriate labels must be keyed to the content—and, without knowing the content, it is hard to be more specific. Generally speaking, it is likely that FOUO information would be relabeled as CUI; whether it would be labeled as Basic or Specified would depend on which type of CUI is involved and whether there is law or regulation directing special handling instructions.

The last three entries in Table 2.3 show legacy DoD labeling practices that are nearly or almost in compliance with the new NARA guidelines. The categories for PROPIN (which generally relates to information generated by DoD business partners and shared with DoD on a sensitive basis) and Source Selection Sensitive (which relates to acquisition and procurement matters leading to award of a contract) are reflected in the categorical and subcategorical structure in the CUI Registry. That means that DoD must only change its practices to label such material appropriately (though the identification of such material as CUI is appropriately being managed). Here, compliance must come in the form of the labeling protocol in ISOO's publication *Marking Controlled Unclassified Information*.[30]

The final label category—TDS—is incorporated into the new CUI Registry as the LDCMs. These control markings closely track the previous markings allowed in DoDM 5200.01, but the way the labels are to be applied has changed under the NARA program.

In summary, the crosswalking of the legacy DoD labels with the new NARA labels reveals that the legacy labels, though now disallowed, are generally subsumed in the new categories and subcategories of CUI that NARA has identified as part of the new regulation. This

---

[29] DoDM 5200.01, 2012.

[30] NARA, 2016c.

is a positive development because all broadly applicable classes of CUI that the DoD acquisition community regularly handles, produces, or acquires are covered by the CUI Registry. The difficult path forward, however, comes from the need to learn the details and definitions of the new categories, identify which are most relevant to the DoD acquisition community, and learn a new labeling regime for CUI.

Public Disclosure and FOIA

CUI labels do not affect the determination about whether information can be released publicly, for example, in response to a FOIA request. As noted above, the CUI scheme is a set of internal administrative controls, largely created because of an asserted need to limit dissemination (and running counter to the original intent of increasing information sharing). Government personnel may be subject to administrative punishment for mishandling properly labeled CUI, but that does not relate to official public disclosure. Further, some CUI markings are based on true legal requirements to protect and limit dissemination of some information, such as source selection information, but the presence of a CUI label does not dictate the outcome of a public release decision.

The procedure for reviewing information for potential public disclosure remains the same and is roughly summarized as follows:

1. Agencies receive a FOIA request and seek out the office that may hold the requested information.
2. Staff in the functional office search agency records.
3. Staff in the functional office make an initial, proposed determination about releasability, including noting which portions (if any) should be withheld under one of the FOIA exemptions.
4. An agency FOIA office may perform a second review of this initial determination.
5. The requester is notified of the agency's decision, which may include the entire requested record, part of the record (with some information redacted), no part of the record, or a statement that the information could not be located.

According to DOJ officials, DOJ provides broad guidance about FOIA across the government and will consult with agency personnel when they have specific questions.

If an agency decides to withhold information requested under FOIA, the requester can sue for release, and DOJ would decide whether to defend the agency's decision in court.

Further, if an agency decides to release information asserted to be PROPIN, the original provider of the information (e.g., a DoD contractor) can initiate a "reverse FOIA" lawsuit to prevent the information's release.

This sequence of events proceeds without regard to CUI labels, and it is possible that a document that is properly labeled as CUI will ultimately be released in full in response to a FOIA request. Whereas FOUO was putatively linked to the FOIA exemptions (but still did not affect the review or release decision), the CUI scheme has no relationship to FOIA and public disclosure. The only consideration is whether the information requested for public release falls within one of the FOIA exemptions, not the CUI Registry.

DoD Implementation Is Still Over the Horizon

In light of the CUI reform effort's shift toward centralization and standardization of the labels for CUI, the biggest concern ultimately is that DoD employees and contractors must now learn the new categories, subcategories, labels, and LDCMs, and unlearn the previous markings and previously established ways of handling CUI material. This chapter has provided an overview of what NARA requires, but many details are still evolving or have yet to be established.

However, the review of the new CUI markings conducted herein—particularly the cross-walk of past DoD acquisition CUI markings with the NARA CUI markings—indicates that there is a high degree of overlap in the content, if not the nomenclature, of the labels. This suggests that controlling CUI will continue at a robust level. The transition period for converting practice (and legacy materials) to the new CUI Registry standard will be uncomfortable, as any such changes are. There are not, however, glaring gaps in what AT&L wishes to control, in what NARA deems CUI, or in who is responsible at the agency level for establishing efficient policy for the future.

Still, there will be trouble in translating policy into practice. Even detailed, clear policy can fail to produce consistent results when it comes to labeling material like CUI. In DoD, for example, FOUO was, by policy, tied explicitly to FOIA exemptions. This connection created relatively well-defined boundaries for what was and was not required to be marked. But as Riposo et al. and McKernan et al. presented in previous RAND studies, significant confusion and misperception about FOUO, in particular, exists among DoD personnel, despite the fact that official DoD policies were clear.

CHAPTER THREE

Overview of Aggregation of Acquisition Information

The second major piece of this analysis is on the compilation or aggregation of acquisition information, which is a risk that needs to be mitigated by those in DoD who are information managers for large amounts of centralized data. This chapter presents current policy on aggregation and compilation, provides a potential framework for managers to consider in addressing aggregation concerns in DoD acquisition information systems, and discusses implications for CUI.

Benefits of Aggregating Acquisition Information

As demonstrated in McKernan et al., 2017, DoD holds a tremendous amount of acquisition data, but those data are of limited value in the disparate and unstructured forms in which they often are stored. Also, many data sets are built for compliance and reporting of individual acquisition programs, not for analytic efforts. In 2013, USD(AT&L) began issuing a series of annual reports on the performance of the Defense Acquisition System.[1] These reports provided the results of analyzing several DoD data sources, including manually aggregating data in some instances. These reports, along with the growing trend of so-called "big data" tools, exemplify the potential value of aggregating more acquisition data, namely that significant management issues may not be discovered by examining programs one at a time. Further, data aggregation helps create a historical record to draw on for program execution and new program development.

Policies on Aggregation Creating Classified Information

Integrating DoD acquisition data provides known benefits, yet officials need to remain aware of the potential to expose classified information through data aggregation. As stated in the DoD Information Security Program Manual,

---

[1] See USD(AT&L), *Performance of the Defense Acquisition System: 2013 Annual Report*, Washington, D.C.: U.S. Department of Defense, June 28, 2013; Under Secretary of Defense, Acquisition, Technology, and Logistics, *Performance of the Defense Acquisition System: 2014 Annual Report*, Washington, D.C.: U.S. Department of Defense, June 13, 2014; USD(AT&L), *Performance of the Defense Acquisition System: 2015 Annual Report*, Washington, D.C.: U.S. Department of Defense, September 16, 2015; USD(AT&L), *Performance of the Defense Acquisition System: 2016 Annual Report*, Washington, D.C.: U.S. Department of Defense, October 24, 2016.

Search capabilities and data mining tools make discovery and correlation of available information fast and simple. This ability to discover and analyze militarily relevant data creates the need to pay particular attention to classified compilations of data elements.[2]

Although the existing literature on the subject describes data "aggregation" or "compilation" in a few different ways, a helpful definition of the term is the "compilation of individual data systems and data that could result in the totality of the information being classified, or [could be] classified at a higher level, or [could be] of beneficial use to an adversary."[3] For the sake of clarity, U.S. government agencies utilize the terms *data aggregation* and *data compilation* interchangeably in reference to this trend. While commercial-sector literature on the risk of data aggregation is limited, the subject is addressed by federal law and agency-specific policy guidance.

Federal Policy

The only known government-wide policy concerning data aggregation is contained in EO 13526. Among several provisions, the order states that

compilations of items of information that are individually unclassified may be classified if the compiled information reveals an additional association or relationship that: (1) meets the standards for classification under this order; and (2) is not otherwise revealed in the individual items of information.[4]

Within the classification standards listed in Section 1.1 of the order, the following condition is particularly pertinent to data aggregation:

the original classification authority determines that the unauthorized disclosure of the information reasonably could be expected to result in damage to the national security, which includes defense against transnational terrorism, and the original classification authority is able to identify or describe the damage.[5]

This language places the responsibility for compilation risks squarely on the original classification authority (OCA), who is the subject-matter expert on a particular program. Note also that while the EO language appears broad, the OCA must show both a reasonable expectation of damage from aggregation *and* what that damage would be. In other words, an OCA must have adequate justification in order to withhold information, and the only consideration is whether the aggregation creates classified information.

In 2010, the ISOO published guidance to assist in implementing EO 13526 across the government in the form of a directive codified in 32 CFR Parts 2001 and 2003. The ISOO directive states:

---

[2]   DoDM 5200.01, *DoD Information Security Program: Overview, Classification, and Declassification*, Vol. 1, February 24, 2012, p. 70.

[3]   Committee on National Security Systems, *National Information Assurance Glossary*, Fort Meade, Md., Instruction No. 4009, April 26, 2010.

[4]   EO 13526, 2009, p. 711.

[5]   EO 13526, 2009, p. 707.

1. Any determination that unclassified information is classified through compilation is a *derivative* classification action based upon existing original classification guidance.
2. Cases of potential classification by aggregation will be referred to the OCA with jurisdiction over the data to make an original classification decision.
3. If the compiled information is determined by the OCA to be classified, clear instructions must appear with the compiled information to indicate which individual portions constitute a classified compilation.[6]

The OCAs create the original classification guidance, review individual data elements for classification issues, and make clear which combinations are classified. The OCAs relevant to acquisition data are likely the individual acquisition program managers, and as a receiver and handler of acquisition data, the Deputy Director, ARA/EI within OUSD(AT&L), must rely on the classification guidance when considering whether data can be aggregated and remain unclassified.

DoD, the Department of Commerce, and the Department of Homeland Security (DHS) have all released policy guidance on classification by compilation, which will be explicated below.

DoD Policy

DoD addresses potential classification by data aggregation in four distinct policy documents, which all draw from the ISOO directive.[7] The policies generally state that DoD needs to be aware of the potential for aggregation, and DoDM 5200.01 has the most-extensive guidance on this topic. The manual contains specific instruction regarding the decisionmaking process involved with classification by compilation and, as provided in the ISOO directive, places responsibility for a final decision with the OCA who has purview over the program that creates or generates the compilation issue.[8] This manual also prescribes portion-marking requirements for each data element, when applicable, so that when disaggregated, the classification of each individual element can be determined.[9] Lastly, the issuance recommends consistently withholding specified data elements from public Internet posting to diminish the opportunity for others to create the classified compilation.[10]

Relative to ARA/EI and acquisition information systems, the "program manager or other official responsible for the database, application, or program that creates or generates the compilation is responsible for facilitating, as necessary, a security classification review with other appropriate OCAs for the constituent items of information."[11] To that end, ARA/EI is writing an access and dissemination control handbook for the information systems that it manages, which will address the guidance above. Any specific combinations of unclassified data elements

---

[6] NARA, 2016a.

[7] Classification by compilation is addressed in DoDM 3020.45-M, *Defense Critical Infrastructure Program Security Classification Manual*, Vol. 3, Washington, D.C., February 15, 2011; DoDM 5200.01, 2012; DoDM 5205.02-M, *DoD Operations Security Program Manual*, Washington, D.C., November 3, 2008; and DoD Instruction 8550.01, *DoD Internet Services and Internet-Based Capabilities*, Washington, D.C., September 11, 2012.

[8] DoDM 5200.01, 2012, p. 42.

[9] DoDM 5200.01, 2012, p. 43.

[10] DoDM 5200.01, 2012, p. 43.

[11] DoDM 5200.01, 2012, p. 42.

that are deemed to be classified based on existing classification guidance from OCAs should be documented in this handbook.

There is also important policy guidance on classification by compilation contained in DoDM 3020.45-M on the Defense Critical Infrastructure Program. The manual states that in the circumstance where a holder of information has reason to believe that a compilation of unclassified data should be classified, then the information should be marked with the anticipated level of classification and with the notation *Pending Classification Review*. Thereafter, the issue should be transmitted to the Deputy CIO for a final classification determination within 60 days of receipt of the request to review.[12] The DoD Operations Security Manual (DoDM 5205.02-M) importantly provides instruction to the DoD components concerning data aggregation. The manual states that the Army, Navy, and Air Force should review those information systems designed for net-centric interoperability for potential classification by compilation issues. Critically, these system owners should address data aggregation issues during the initial planning stages and provide guidance on mitigation strategies.[13] In the context of ARA/EI's management of acquisition data submitted by myriad programs across all DoD components, this language is vital because it also places responsibility on the DoD components for identifying and mitigating potential issues surrounding classification by compilation.

U.S. Computer Emergency Readiness Team Policy

The U.S. Computer Emergency Readiness Team (US-CERT) is a unit within the DHS that publishes actionable information for federal agencies to protect against cyberattacks.[14] US-CERT also provides guidance on protecting aggregated data against the evolving cyber threat. They particularly emphasize the importance of good *management principles*, stating that "a good set of commonly accepted management principles aids an organization's leaders in determining what protection strategies are best applied to secure aggregated data," including accountability, adequacy, awareness, compliance, measurement, response, and risk management.[15] US-CERT also highlights the significance of good *security practices*, maintaining that "to be effective and of greatest value, [security] practices should guide control selection and address risk mitigation efforts necessary to adequately protect sensitive aggregated data."[16] Applicable security practice areas include information security strategy, security architecture, incident management, partner management, contingency planning, and disaster recovery.

National Institute of Standards and Technology Policy Offers a Framework to Consider

The National Institute of Standards and Technology (NIST) provides a framework for categorizing information systems that may breed classification by aggregation issues. NIST Special Publication 800-60 delivers specific guidance on aggregation, stating that if

> [a] review reveals increased sensitivity or criticality associated with information aggregates, then the system security objective impact levels may need to be adjusted to a higher level

---

[12] DoDM 3020.45-M, 2011, p. 9.

[13] DoDM 5205.02-M, 2008, p. 31.

[14] U.S. Computer Emergency Readiness Team, "About Us: Our Mission," webpage, undated.

[15] U.S. Computer Emergency Readiness Team, *Protecting Aggregated Data*, December 5, 2005, p. 12.

[16] U.S. Computer Emergency Readiness Team, 2005, p. 15.

than would be indicated by the security impact levels associated with any individual information type.[17]

This publication adopts Federal Information Processing Standard Publication 199 standards for the potential levels of *security impact* (low, moderate, and high).[18] These impact levels are associated with three stated *security objectives* (confidentiality, integrity, and availability), which are elucidated in Table 3.1.[19]

With security impact and objective defined, NIST Special Publication 800-60 recommends determining an appropriate *security category* for each information type within a system, which merely requires matching the potential security impact (low, moderate, or high) for each applicable security objective. The following notation is recommended for marking these data elements:

Security Category information type = {(confidentiality, impact), (integrity, impact), (availability, impact)}.[20]

An example of an implementation of this guidance is provided in NIST Special Publication 800-60, Volume I Revision:

EXAMPLE 1: An organization managing public information on its web server determines that there is no potential impact from a loss of confidentiality (i.e., confidentiality requirements are not applicable), a moderate potential impact from a loss of integrity, and a moderate potential impact from a loss of availability. The resulting security category of this information type is expressed as:

Table 3.1
Special Publication 800-60 Information and Information System Security Objectives

Security Objectives	Federal Information Security Management Act Definition (44 U.S.C., Sec. 3542)	Federal Information Processing Standard 199 Definition
Confidentiality	"Preserving authorized restrictions on information access and disclosure, including means for protecting personal privacy and proprietary information . . ."	A loss of *confidentiality* is the unauthorized disclosure of information.
Integrity	"Guarding against improper information modification or destruction, and includes ensuring information non-repudiation and authenticity . . ."	A loss of *integrity* is the unauthorized modification or destruction of information.
Availability	"Ensuring timely and reliable access to and use of information . . ."	A loss of *availability* is the disruption of access to or use of information or an information system.

SOURCE: NIST, 2008, p. 9.

---

[17] NIST, *Guide for Mapping Types of Information and Information Systems to Security Categories*, Vol. 1, Special Publication 800-60, August 2008, p. 27.

[18] Federal Information Processing Standard Publication 199 is a compulsory security standard, as mandated by the Federal Information Security Management Act.

[19] NIST, 2008, p. 9.

[20] NIST, 2008, p. 11.

Security Category public information = {(confidentiality, N/A), (integrity, moderate), (availability, moderate)}.[21]

Such a data profiling process, as illustrated in Figure 3.1, can be incorporated into ARA/EI's information management practices to mitigate any potential classification by compilation issues. This framework can allow ARA/EI to examine potential classification issues.

Postulated Security Concerns for Unclassified Aggregation

One question posed throughout this study is whether data aggregation creates potential security concerns. As described above, there may be instances in which combining pieces of otherwise unclassified data creates a product that is deemed classified, which is the best example of how aggregation causes a security concern. Our discussions with government stakeholders revealed some concern about the potential of aggregation of acquisition information, particularly through information systems that store large amounts of information. The consensus in the discussions was that it is difficult to discuss and determine when aggregation occurs in practice unless a concrete example is used. Moreover, there are few instances of DoD acquisition data being part of a publicly available, Internet-accessible database, and OCAs are careful about what information is posted on the open web. Ultimately, DoD information managers need to remain diligent about the information under their control.

Aggregation of Public Information

Data released in response to a FOIA request cannot be reclaimed by the government and retroactively found exempt from public disclosure. This is partly legal and partly practical. Once a FOIA release is made, the information remains public, according to DOJ officials. Further,

Figure 3.1
Special Publication 800-60 Security Categorization Process Execution

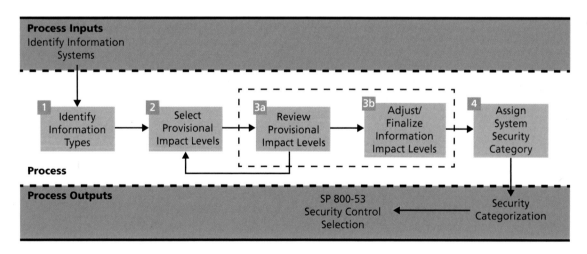

SOURCE: NIST, 2008, p. 12.
RAND RR2221-3.1

---

[21] NIST, 2008, p. 22.

in the modern era of electronic data storage and archiving, there is no practical way to prevent aggregation of public information once it has been released. Consequently, DoD cannot take any steps relative to that information, even if they believe aggregation techniques somehow change the proper treatment of that information. DoD officials could attempt to withhold that type of information in the future, but making such a legal argument would be difficult. Further, as discussed above, simply claiming that information is now CUI does not affect the treatment of that information under FOIA.

Aggregation of CUI

The CUI regime is a binary system: information is either CUI or it is not. Consequently, aggregated sets of CUI simply carry any special handling required of the separate pieces. Given the large number of CUI categories, it may be possible that two sets of CUI Basic somehow become CUI Specified and require different handling. However, this different handling does not make the information "more CUI;" it remains under the administrative CUI control scheme. This is different than derivative classification, in which aggregating information could require a fundamental change in the electronic systems in which the information can be stored (e.g., an unclassified network versus a classified network). A required shift from CUI Basic to CUI Specified would likely be relatively apparent based on the original, disaggregated information, but the new information could still be stored and handled on the same unclassified electronic systems.

Aggregation of Public and Sensitive Information

Similar to aggregating multiple pieces of CUI, aggregating CUI with public data creates a data set that remains CUI based on the handling requirements of the original data.

A scenario that requires attention is whether combining uncontrolled, nonpublic data creates a data set that requires CUI marking and handling. A common example of this would be combining deidentified personnel information with personally identifiable information (e.g., names matched up with Social Security Numbers).

Acquisition programs—especially Major Defense Acquisition Programs (MDAPs)—generally submit complete, detailed, discrete information to a central information system. This information may be deidentified (e.g., removing company names) or summarized (sometimes referred to as "aggregation") for analytic purposes in ways to protect potentially proprietary information, like profit data on DoD contracts. For example, the USD(AT&L) reports on the performance of the Defense Acquisition System included results of exactly that type. In other words, acquisition data sets already contain all of the information that would require CUI marking and handling.

Nevertheless, it is the responsibility of data submitters and data holders to understand whether their information must be treated as CUI, similar to OCAs and creating classified information by combining unclassified data. Submitters of data (sometimes referred to as "owners") will have to determine which data elements need to be treated as CUI, and the information system manager (like ARA/EI) will have to ensure proper handling based on those determinations. A situation could theoretically arise in which combining multiple sets of acquisition data reveals potentially proprietary information. The most likely situation is that CUI will be added to a larger set of uncontrolled data, requiring the addition of business rules to ensure proper access and handling procedures. As noted above, CUI is more expansive than FOUO treatment, but it is a simple binary: controlled or not.

Disaggregation Is Also a Potential Challenge

During our discussions with DoD subject-matter experts, we discussed aggregation of acquisition information; however, the concept of disaggregation of information was also raised. As part of the discussions, it was noted that it is difficult to identify when information is no longer CUI. For example, a prime contractor may have multiple parts that it needs to collect from its suppliers. Individually, information (e.g., a particular size of bolt) may not be CUI, but when parts are considered together, the result may be CUI. The prime contractor might find it difficult to determine when to stop marking CUI as disaggregation to various suppliers occurs. A policy or a framework for understanding disaggregation can help relieve marking requirement pressures, but as with aggregation, this requires definitive understanding of the labels on all of the individual pieces of information. This illustrates one of the benefits of using portion marking on pieces of information and also illustrates the need to reject information that lacks markings altogether.

CHAPTER FOUR

Conclusions and Options

In this analysis on the CUI reform effort and in particular, the aggregation of CUI, we were able to develop a set of findings and some options for OUSD(AT&L) to consider, which are provided below.

Implementation of CUI Reform Will Be Significant for DoD

Implementation of the new CUI program is destined to be disruptive. The new protocols negate and replace what have become customary, perhaps intractable CUI labeling practices. Notably, the FOUO label, widely used on DoD documentation, is now prohibited. Its replacement sometimes requires a more intensive process for choosing an appropriate label closely keyed to the specific content of the information being labeled (CUI Specified). Furthermore, legacy materials—those that may need to be relabeled under the new regime—will further enhance the disruption in DoD practices. That relabeling process of legacy information could require contacting original owners of the information, or the creation of a new adjudication process—a time-intensive effort, to say the least. When compared with the precursor use of labels on CUI material (like FOUO), the new process is much more descriptive and prescriptive.

Given the emphasis on the importance and specificity of labeling information, **training is likely to be extensive for both DoD employees and contractors.** Every DoD employee and contractor will need to receive training on the new marking schemas. If DoD decides to implement a portion-marking requirement, training will be more complicated, as not all DoD employees and contractors are familiar with the portion-marking process.

Implementation is currently unfunded, and it is not clear how much of a financial burden it will be on those who need to implement. In particular, implementation will require potentially significant changes to information systems, depending on how easy it will be to integrate new markings into pre-existing materials. This may be particularly difficult given that not all structured and unstructured data have markings in the current CUI regime.

Several commonly used labels on acquisition information are no longer permitted, which will leave DoD employees and contractors looking for the next "FOUO." In Chapter Two, we conducted a crosswalk to better understand which labels would be eliminated. In the post-CUI regulation era, "Business Sensitive," "Competition Sensitive," "For Official Use Only," and "Pre-Decisional" are largely noncompliant, as only NARA-approved banner and portion markings can be used on CUI. In response, DoD will have to redesign its labeling processes to match the CUI registry approved labels. The crosswalking of the legacy DoD labels with the new NARA labels reveals that the legacy labels, though now disallowed, are generally subsumed in the new categories and subcategories of CUI that NARA has identified as part of the new regulation. This is a positive development, as it means that all broadly applicable classes of CUI that the DoD acquisition community regularly handles, produces, or acquires are covered by the CUI Registry. The difficulty comes from the need to learn the details and definitions of the new categories, identify which are most relevant to the DoD acquisition community, and learn a new labeling regime for CUI.

CUI was originally meant to increase sharing, but recent CUI efforts have taken on the tenor of a security protocol—which may inadvertently exacerbate sharing problems. This new process has parallels in the classified and information security protocols, including similar banner marking requirements (and possible portion-marking requirements) as well as enforcement mechanisms for noncompliance. Each agency may also have its own office to record and collect security incidents and audit materials. All of this will require additional security reviews and personnel.

A Still-Unclear Path Forward

In light of the CUI reform effort's shift toward centralization and standardization of CUI labels, the biggest concern ultimately appears to be the process of learning the new categories, subcategories, labels, and LDCMs. We provided an overview of NARA requirements, but many details are still evolving or have yet to be established. Equally important is the cultural and practical deprogramming of the previous markings and established ways within DoD of handling CUI material.

However, the review of the new CUI Registry markings conducted herein indicates a high degree of overlap in the content, if not the nomenclature, of past and present CUI labels. This suggests that the major thrust of controlling CUI is going to continue at a robust level. The transition period for converting to the new CUI Registry standard will be uncomfortable, as such changes often are. There are not, however, glaring gaps in what OUSD(AT&L) wishes to control, in what NARA deems CUI, or in who is responsible at the agency level for establishing efficient policy for the future.

Still, there will be trouble in translating policy into practice. Even detailed, clear policy can fail to produce consistent results when it comes to labeling material. For example, in DoD, FOUO was, by policy, tied explicitly to FOIA exemptions. This connection created relatively well-defined boundaries for what was and was not required to be marked. But as presented in previous research, significant confusion and misperception about FOUO, in particular, exists among DoD personnel, despite the fact that official DoD policies were clear.[1]

Ultimately, it will be important for agencies within the federal government to closely monitor the implementation of this effort. It is possible that the best intentions here will not produce the benefits that have been originally touted, given that this is a massive effort that may be implemented unevenly across the government and shifted in overall approach over time to a security posture over an information-sharing posture.

---

[1]   Riposo et al., 2015, McKernan et al., 2016; and McKernan et al., 2017.

Aggregation of Acquisition Information

The second major piece of this analysis is on the aggregation of CUI. We reviewed how aggregation or compilation is present in policy, the benefits of aggregation, and the security concerns. We focused on aggregation as part of this analysis on CUI given the Deputy Director, ARA/EI's role as an information manager of acquisition data.

The benefits to acquisition decisionmakers and personnel of the compilation of acquisition information cannot be overstated. One important example is that aggregation allows key decisionmakers to provide the "state of the acquisition portfolio" to Congress. **However, DoD is also concerned about adversaries accessing aggregated information or aggregating information themselves.**

Generally, policy and guidance on aggregation advises DoD and others within the federal government to be aware of the potential for aggregation. In practice, information managers have several major concerns involving aggregation:

- The process of examining all potential combinations of information that could result in aggregation is a significant effort given tight resources and the presence of information that does not have any markings.
- Also, through our discussions with subject-matter experts, we found that it is hard to understand or pinpoint what aggregation is without concrete examples.
- Finally, information managers are concerned about the prevalence of easily accessible "big data" tools to aggregate information. The security possibilities can be seemingly endless to consider as a steward of acquisition information.

We addressed some of those concerns in Chapter Three. It is well understood that combining pieces of otherwise unclassified data may create a product that is deemed classified, which is the best example of how aggregation causes a security concern. Our discussions with government stakeholders revealed some concern about the potential of aggregation of acquisition information, particularly through the information systems that store large amounts of information; however, the consensus in the discussions was that it is difficult to determine when aggregation occurs in practice unless a concrete example is used.

Specific Examples of Aggregation

There are three specific examples of aggregation that we assessed: aggregation of public information, aggregation of CUI, and aggregation of public information and CUI. Data or information that has gone through the process to be publicly released is available to the public. Once made public, the information remains public, according to DOJ officials. Further, in the modern era of electronic data storage and archiving, there is no practical way to prevent aggregation of public information once it has been released. If there are concerns regarding the release of public information because of potential aggregation, they would need to be expressed during the public release process.

The CUI regime is a binary system: information is either CUI or it is not. Consequently, aggregated sets of CUI simply carry any special handling required of the separate pieces. Given the large number of CUI categories, it may be possible that two sets of CUI Basic, once aggregated, would somehow become CUI Specified and would require different handling. However, this different handling does not make the information "more CUI;" it remains under the administrative control scheme of CUI. This is different than derivative classification, where aggregating information could require a fundamental change in the electronic systems on which the information can be stored (e.g., an unclassified network [Nonsecure Internet Protocol Router Network, or NIPRNet]) versus a classified network [Secure Internet Protocol Router Network or SIPRNet]). A required shift from CUI Basic to CUI Specified would likely be relatively apparent based on the original, disaggregated information, but could still be stored and handled on the same unclassified electronic systems.

Aggregating CUI data with public data could result in a different CUI determination. To this end, it is the responsibility of data submitters and data holders to understand whether the aggregated data must be treated as CUI. Submitters of data (sometimes referred to as "owners") will have to determine which data elements need to be treated as CUI, and the information system manager (like ARA/EI) will have to ensure proper handling based on those determinations.

During our discussions, concerns regarding disaggregation of information were also raised. It was noted that it is difficult to identify when information is no longer CUI. For example, a prime contractor may have multiple parts that it needs to collect from its suppliers. When the parts are considered together, the result may be CUI, but individual parts may not be CUI. Policy or a framework for understanding disaggregation can help relieve marking requirement pressures, but as with aggregation, this requires definitive understanding of the labels on all of the individual pieces of information which would require portion-marking on pieces of information and also the need to reject information that lacks markings.

Options for OUSD(AT&L) to Consider in Regard to CUI Reform and Aggregation

We offer some options for OUSD(AT&L) to consider regarding both the CUI reform effort and aggregation. As it currently stands, CUI reform will have significant effects on the management and handling of acquisition data. Lack of participation could lead to major challenges in OUSD(AT&L)'s day-to-day operations, so OUSD(AT&L) should

- **Identify a point of contact to help advise and transition OUSD(AT&L) to the new marking regime.** This function could be assigned to someone who already has a role in the organization. For example, this role could be assigned to a senior member of the staff as a collateral duty during the transition.
- **Actively engage in discussions with USD(I)** because USD(I) will benefit from understanding OUSD(AT&L)'s needs while revising DoDM 5200.1, Vol. 4. OUSD(AT&L)'s participation in reforms to date has been limited to mostly private-sector and technical information.
- **Work closely with NARA as needed** to understand some of the current guidance that has been issued at the federal level.
- **Hold small working groups with the military services and DoD functions** (e.g., Comptroller) in order to further understand the implications of this effort.
- **Begin to work to identify training resource requirements.** AT&L may want to create its own focused training for the CUI categories that its staff are most likely to use rather than rely solely on the broader DoD implementation training.

- **Wait to implement until USD(I) completes the guidance** per USD(I)'s strong recommendation. Several key pieces of implementation (e.g., portion marking) are still being discussed.
- **Carefully monitor changes** to both the CUI registry and any potential changes to the overall federal CUI strategy by the Trump administration.
- In regard to data aggregation, **Deputy Director, ARA/EI in OUSD(AT&L) should consider using NIST's aggregation tool** described in Chapter Three as a mechanism for systematically combing through information systems for potential aggregation.

Overview of NARA Categories of Importance to the DoD Acquisition Community

In Chapter Two, we provided the NARA CUI categories that are most relevant to defense acquisition professionals. They are:

- Controlled Technical Information
- Critical Infrastructure
- Financial
- Information Systems Vulnerability Information
- Privacy
- Procurement and Acquisition
- Proprietary Business Information
- Nuclear
- Export Control.

In this appendix, we provide additional information on those categories that may be useful as acquisition professionals start to understand how this reform will affect day-to-day operations. Because the key effort for information managers will be to understand what information triggers the control and labeling regime, we review each category; the definition of the information under its purview; the labels NARA requires; and, where necessary, areas of ambiguity associated with each.

Controlled Technical Information

The category of Controlled Technical Information (CTI) (NARA label: CUI//SP-CTI) is a CUI Specified category of information relating to technical information. CTI is defined as "technical information with military or space application that is subject to controls on the access, use, reproduction, modification, performance, display, release, disclosure, or dissemination," and examples of CTI include

> research and engineering data, engineering drawings, and associated lists, specifications, standards, process sheets, manuals, technical reports, technical orders, catalog-item identifications, data sets, studies and analyses and related information, and computer software executable code and source code.[1]

---

[1]   NARA, "CUI Registry: Controlled Technical Information," webpage, last reviewed July 20, 2017f.

The definition also references the Defense Federal Acquisition Regulation Supplement (DFARS) clauses relating to technical information.[2] NARA's definition of CTI also includes instruction on handling CTI, explicitly stating that TDSs must be included in the markings. At the moment, NARA's definition requires one of the distributions statements outlined in DoDI 5230.24—although NARA's own LDCMs seem to preclude the agency's own dissemination-marking regime.[3] This is an area that is presently unresolved.

The effect of the CTI category and labeling regime is minimal because the category is, by design, defined by existing DoD policy and the DFARS. That assumes, however, that the preexisting labeling regime for technical data was being handled appropriately. Questions and ambiguities in application of the label—and for every label—should be directed to USD(I).

Critical Infrastructure

The Critical Infrastructure category (NARA label: CUI or CUI//CRIT) is a CUI Basic category; therefore, the owner of the information may choose whether to label it with the generic CUI banner or with the more-specific CUI//CRIT label. NARA defines Critical Infrastructure as

> Systems and assets, whether physical or virtual, so vital that the incapacity or destruction of such may have a debilitating impact on the security, economy, public health or safety, environment, or any combination of these matters, across any Federal, State, regional, territorial, or local jurisdiction.[4]

Within this broad category, however, there is a subcategory of CUI Basic called DoD Critical Infrastructure Security Information. This is not a CUI Specified category because there are not explicit handling instructions in law or regulation. However, DoD information managers should know that this subcategory exists specifically for DoD. The subcategory is CUI or CUI//DCRIT (DoD Critical Infrastructure Security Information).

The DCRIT subcategory is further defined as:

> Information that, if disclosed, would reveal vulnerabilities in the DoD critical infrastructure and, if exploited, would likely result in the significant disruption, destruction, or damage of or to DoD operations, property, or facilities, including information regarding the securing and safeguarding of explosives, hazardous chemicals, or pipelines, related to critical infrastructure or protected systems owned or operated on behalf of the DoD, including vulnerability assessments prepared by or on behalf of the DoD, explosives safety information (including storage and handling), and other site-specific information on or relating to installation security.[5]

---

[2] Code of Federal Regulations, Title 48, Section 9.505-4, Obtaining Access to Proprietary Information, October 1, 2002.

[3] Department of Defense Instruction 5230.24, *Distribution Statements on Technical Documents*, Washington, D.C., August 23, 2012.

[4] NARA, "CUI Registry: Critical Infrastructure," webpage, last reviewed November 3, 2016b.

[5] NARA, 2017a.

In practice, DoD acquisition professionals are rarely expected to handle this category of information, but they will be expected to be aware of the proper use, handling, and labeling associated with this type of CUI because it is specifically identified within the CUI Registry structure as a DoD-relevant subcategory.

Financial

The Financial category of CUI is defined as information

> [r]elated to the duties, transactions, or otherwise falling under the purview of financial institutions or United States Government fiscal functions. Uses may include, but are not limited to, customer information held by a financial institution.[6]

> Financial CUI may be either Basic or Specified, depending on the content and whether it falls within a legal or regulatory framework for the handling of the information. Hence, the label may be CUI, CUI//FNC, or CUI//SP-FNC, as circumstances warrant. Note that Financial CUI is a distinct category from Procurement and Acquisition CUI, which is discussed below.

Within the broader category, there is a subcategory of CUI Specified Financial information that requires special handling and special labeling. That is Budget information (labeled as CUI//SP-BUDG), defined as "information concerning the federal budget, including authorizations and estimates of income and expenditures."[7]

Information Systems Vulnerability Information

This category of CUI Basic concerns

> information that if not protected, could result in adverse effects to information systems. Information system means a discrete set of information resources organized for the collection, processing, maintenance, use, sharing, dissemination, or disposition of information.[8]

Information falling within this category should be labeled CUI or CUI//ISVI.

This category of CUI may be particularly pertinent given the data architecture at OUSD(AT&L), and particularly, the efforts of ARA/EI office within OUSD(AT&L). ARA/EI is the manager of several information systems that centralize acquisition data within DoD. AIR, DAMIR, and the Defense Acquisition Visibility Environment (DAVE) are three information systems frequently used for reporting, oversight, and analysis in defense acquisition. AIR provides one central location for all MDAP and Major Automated Information System acquisition documents to support oversight and decisionmaking.[9] DAMIR fulfills several key

---

[6] NARA, "CUI Registry: Financial," webpage, last reviewed July 26, 2017h.

[7] NARA, "CUI Registry: Financial-Budget," webpage, last reviewed August 2, 2017m.

[8] NARA, "CUI Registry: Information Systems Vulnerability Information, webpage, last reviewed July 19, 2017e.

[9] AIR is a repository that contains specific program documents (reports, certifications) used to inform acquisition decisionmaking and oversight.

functions, including reporting, storage, quality assurance, analysis, and oversight; it also tracks the cost, schedule, and performance of major acquisition programs.[10] DAVE also provides acquisition information and support for oversight, analysis, and decisionmaking, including data opportunities that defense analysts can utilize along with some centralized policy and information on other reporting capabilities.

The efforts within OUSD(AT&L) clearly concern the collection, processing, use, sharing, and dissemination of information across many critical stakeholders. Information about their own systems, as well as information about other systems at DoD, may reasonably fall within this category.

Privacy

The Privacy category of CUI has become increasingly familiar to users of personal data (including social security numbers, health records, and financial details). For NARA, the Privacy category (NARA label: CUI or CUI//PRVCY) is a general category that "[r]efers to personal information, or, in some cases, 'personally identifiable information,' as defined in OMB M-17-12, or 'means of identification,' as defined in 18 USC 1028(d)(7)."[11] Subcategories of Privacy CUI include death records, genetic and health information, personnel records of agency employees, and the identity of a person making a report to the inspector general of an agency. Additionally, there is a Military subcategory (still CUI Basic) covering the personally identifiable information of "[a]ny member or former member of the armed forces or affiliated organization of the Department of Defense."[12]

For acquisition professionals, the effect of the Privacy category is tangential to the technical data and procurement information they regularly process. If, however, there is an issue when personally identifiable information is part of the process of obtaining access to acquisition data repositories (as when a Common Access Card is required), then there can be implications for data labeling and handling. At the moment, the indication from NARA seems to be that if Privacy information is in the document, dataset, or data repository, then it must be labeled as such.

Procurement and Acquisition

Procurement and Acquisition information (NARA label: CUI, CUI//PROCURE, or CUI//SP-PROCURE, as appropriate) is "[m]aterial and information relating to, or associated with, the acquisition and procurement of goods and services, including but not limited to, cost or pricing data, contract information, indirect costs and direct labor rates."[13] There are sub-

---

[10] DAMIR has both unclassified and classified versions. It supports the generation, distribution, and archiving of Selected Acquisition Reports, as well as information supporting the Defense Acquisition Executive Summary process. It also includes higher-level earned value management data. Unlike AIR, DAMIR is structured data that users can combine and analyze in multiple ways, serving multiple functions.

[11] NARA, "CUI Registry: Privacy," webpage, last reviewed July 26, 2017i.

[12] NARA, "CUI Registry: Privacy—Military," webpage, last reviewed July 11, 2017c.

[13] NARA, 2017k.

categories for information relating to the Small Business Innovation Research Program and the Small Business Technology Transfer Program. Additionally, there are both CUI Basic and CUI Specified labels relating to Source Selection Information (NARA label: CUI, CUI//SSEL, or CUI//SP-SSEL, as appropriate). NARA's description of the Source Selection Information subcategory reads:

> Per FAR 2.101: any of the following information that is prepared for use by an agency for the purpose of evaluating a bid or proposal to enter into an agency procurement contract, if that information has not been previously made available to the public or disclosed publicly: (Items 1-10).[14]

While NARA guidelines allow for most of this information to be labeled merely as CUI (unless specific law or regulation require the more restrictive CUI Specified labeling protocols to be used), a best practice would be to label information as CUI//PROCURE to better indicate the content of the information subject to CUI safeguards.

Proprietary Business Information

A category likely to be important to DoD acquisition professionals is Proprietary Business Information (NARA Label: CUI, CUI//PROPIN, or CUI//SP-PROPIN, as appropriate). This is

> [m]aterial and information relating to, or associated with, a company's products, business, or activities, including but not limited to financial information; data or statements; trade secrets; product research and development; existing and future product designs and performance specifications.[15]

One issue that has not yet been resolved by NARA concerns the handling instructions for certain types of CUI//SP-PROPIN. Under DoDI 5230.24, proprietary information was supposed to be further labeled with Distribution Statement B, which limited dissemination to U.S. government agencies only.[16] No such dissemination statement or LDCM is incorporated into the CUI Registry. This contrasts with the CTI category, which explicitly incorporates a dissemination label.

Nuclear

As weapon systems have become more advanced, the use of nuclear materials has increased. The CUI Registry contains a category for nuclear-related CUI, described as information related to "protection of nuclear information concerning nuclear reactors, material or security."[17] Labels

---

[14] NARA, "CUI Registry: Procurement and Acquisition—Source Selection," webpage, last reviewed July 25, 2017g.

[15] NARA, "CUI Registry: Proprietary Business Information," webpage, last reviewed July 31, 2017l.

[16] DoDI 5230.24, 2016, p. 17. 48 CFR 9.505-4 relaxes the restriction on U.S. government distribution only, and allows qualified government support contractors, like FFRDCs, to have limited access to proprietary information when in the course of their work for the government.

[17] NARA, "CUI Registry: Nuclear," webpage, last reviewed July 18, 2017d.

for such material could include CUI, CUI//NUC, or CUI//SP-NUC. A special subcategory for defense-related nuclear material exists, "Relating to Department of Defense special nuclear material (SNM), equipment, and facilities, as defined by 32 CFR 223." The subcategory was created to distinguish defense-related programs from Department of Energy programs. Labels for the subcategory would include CUI, CUI//DCNI, or CUI//SP-DCNI.

Export Control

Finally, some divisions within the DoD acquisition community regularly handle cooperation with foreign governments. Involving such matters as cooperative agreements and foreign military sales, such divisions should be aware of an additional CUI category for Export Control. The NARA Registry defines this category as

> unclassified information concerning certain items, commodities, technology, software, or other information whose export could reasonably be expected to adversely affect the United States national security and nonproliferation objectives. To include dual use items; items identified in export administration regulations, international traffic in arms regulations and the munitions list; license applications; and sensitive nuclear technology information.[18]

Labels for the subcategory would include CUI, CUI//EXPT, or CUI//SP-EXPT.

---

[18]   NARA, "CUI Registry: Export Control," webpage, last reviewed July 27, 2017j.

Bibliography

Bush, George W., "Guidelines and Requirements in Support of the Information Sharing Environment," memorandum for the Heads of Executive Departments and Agencies, Washington, D.C., December 16, 2005. As of November 28, 2017:
https://www.gpo.gov/fdsys/pkg/WCPD-2005-12-26/pdf/WCPD-2005-12-26-Pg1874.pdf

———, "Designation and Sharing of Controlled Unclassified Information (CUI)," memorandum for the Heads of Executive Departments and Agencies, Washington, D.C., May 7, 2008. As of November 28, 2017:
https://www.gpo.gov/fdsys/pkg/WCPD-2008-05-12/pdf/WCPD-2008-05-12-Pg673.pdf

Code of Federal Regulations, Title 32, Section 2002, Controlled Unclassified Information, September 14, 2016.

Code of Federal Regulations, Title 32, Section 2002.4(h), Definitions, September 14, 2016.

Code of Federal Regulations, Title 48, Section 9.505-4, Obtaining Access to Proprietary Information, October 1, 2002.

Committee on National Security Systems, *National Information Assurance Glossary*, Fort Meade, Md., Instruction No. 4009, April 26, 2010. As of November 28, 2017:
https://www.ecs.csus.edu/csc/iac/cnssi_4009.pdf

Department of Defense Instruction 5230.24, *Distribution Statements on Technical Documents*, Washington, D.C., August 23, 2012. As of November 29, 2017:
http://www.esd.whs.mil/Portals/54/Documents/DD/issuances/dodi/523027_dodi_2016.pdf

Department of Defense Instruction 8550.01, *DoD Internet Services and Internet-Based Capabilities*, Washington, D.C., September 11, 2012. As of November 28, 2017:
http://www.esd.whs.mil/Portals/54/Documents/DD/issuances/dodi/855001p.pdf

Department of Defense Instruction 8582.01, *Security of Unclassified DoD Information on Non-DoD Information Systems*, Washington, D.C., June 6, 2012.

Department of Defense Manual 3020.45-M, *Defense Critical Infrastructure Program Security Classification Manual*, Vol. 3, Washington, D.C., February 15, 2011.

Department of Defense Manual 5200.01, *DoD Information Security Program: Overview, Classification, and Declassification*, Vol. 1, Washington, D.C., February 24, 2012.

Department of Defense Manual 5200.01, *CUI*, Vol. 4, Washington, D.C., May 12, 2012.

Department of Defense Manual 5205.02-M, *DoD Operations Security Program Manual*, Washington, D.C., November 3, 2008.

DoDI—*See* Department of Defense Instruction.

DoDM—*See* Department of Defense Manual.

EO—*See* Executive Order.

Executive Order 13526, *Classified National Security Information*, Washington, D.C.: The White House, December 29, 2009.

Executive Order 13556, *Controlled Unclassified Information*, Washington, D.C.: The White House, 3 C.F.R. 68675, November 4, 2010.

Knezo, Genevieve J., *CRS Report for Congress: "Sensitive But Unclassified" and Other Federal Security Controls on Scientific and Technical Information: History and Current Controversy*, Washington, D.C.: Congressional Research Service, February 20, 2004. As of September 11, 2017:
https://fas.org/sgp/crs/RL31845.pdf

McKernan, Megan, Nancy Young Moore, Kathryn Connor, Mary E. Chenoweth, Jeffrey A. Drezner, James Dryden, Clifford A. Grammich, Judith D. Mele, Walter T. Nelson, Rebeca Orrie, Douglas Shontz, and Anita Szafran, *Issues with Access to Acquisition Data and Information in the Department of Defense: Doing Data Right in Weapon System Acquisition*, Santa Monica, Calif.: RAND Corporation, RR-1534-OSD, 2017. As of May 10, 2017:
https://www.rand.org/pubs/research_reports/RR1534.html

McKernan, Megan, Jessie Riposo, Jeffrey A. Drezner, Geoffrey McGovern, Douglas Shontz, and Clifford A. Grammich, *Issues with Access to Acquisition Data and Information in the Department of Defense: A Closer Look at the Origins and Implementation of Controlled Unclassified Information Labels and Security Policy*, Santa Monica, Calif.: RAND Corporation, RR-1476-OSD, 2016. As of May 10, 2017:
https://www.rand.org/pubs/research_reports/RR1476.html

NARA—*See* National Archives and Records Administration.

National Archives and Records Administration, "32 CFR Parts 2001 and 2003 Classified National Security Information; Final Rule," webpage, last reviewed August 15, 2016a. As of November 29, 2017:
https://www.archives.gov/isoo/policy-documents/isoo-implementing-directive.html

———, "CUI Registry: Critical Infrastructure," webpage, last reviewed November 3, 2016b. As of November 28, 2017:
https://www.archives.gov/cui/registry/category-detail/critical-infrastructure.html

———, *Marking Controlled Unclassified Information*, Version 1.1, December 6, 2016c. As of May 30, 2017:
https://www.archives.gov/files/cui/20161206-cui-marking-handbook-v1-1.pdf

———, "Chronology," webpage, last reviewed December 14, 2016d. As of November 28, 2017:
https://www.archives.gov/cui/chronology.html

———, "CUI Registry: Critical Infrastructure-DoD Critical Infrastructure Security Information," webpage, last reviewed June 19, 2017a. As of November 28, 2017:
https://www.archives.gov/cui/registry/category-detail/critical-infrastructure-dod-security-info.html

———, "CUI Registry: Limited Dissemination Controls," webpage, last reviewed June 19, 2017b. As of November 29, 2017:
https://www.archives.gov/cui/registry/limited-dissemination

———, "CUI Registry: Privacy—Military," webpage, last reviewed July 11, 2017c. As of November 28, 2017:
https://www.archives.gov/cui/registry/category-detail/privacy-military.html

———, "CUI Registry: Nuclear," webpage, last reviewed July 18, 2017d. As of November 28, 2017:
https://www.archives.gov/cui/registry/category-detail/nuclear.html

———, "CUI Registry: Information Systems Vulnerability Information," webpage, last reviewed July 19, 2017e. As of November 28, 2017:
https://www.archives.gov/cui/registry/category-detail/info-systems-vulnerability-info.html

———, "CUI Registry: Controlled Technical Information," webpage, last reviewed July 20, 2017f. As of November 28, 2017:
https://www.archives.gov/cui/registry/category-detail/controlled-technical-info.html

———, "CUI Registry: Procurement and Acquisition—Source Selection," webpage, last reviewed July 25, 2017g. As of November 28, 2017:
https://www.archives.gov/cui/registry/category-detail/proprietary-source-selection.html

———, "CUI Registry: Financial," webpage, last reviewed July 26, 2017h. As of November 28, 2017:
https://www.archives.gov/cui/registry/category-detail/financial.html

———, "CUI Registry: Privacy," webpage, last reviewed July 26, 2017i. As of November 28, 2017:
https://www.archives.gov/cui/registry/category-detail/privacy.html

———, "CUI Registry: Export Control," webpage, last reviewed July 27, 2017j. As of November 28, 2017:
https://www.archives.gov/cui/registry/category-detail/export-control.html

———, "CUI Registry: Procurement and Acquisition," webpage, last reviewed July 27, 2017k. As of November 28, 2017:
https://www.archives.gov/cui/registry/category-detail/procurement-acquisition.html

———, "CUI Registry: Proprietary Business Information," webpage, last reviewed July 31, 2017l. As of November 28, 2017:
https://www.archives.gov/cui/registry/category-detail/proprietary-business-info.html

———, "CUI Registry: Financial-Budget," webpage, last reviewed August 2, 2017m. As of November 28, 2017:
https://www.archives.gov/cui/registry/category-detail/financial-budget.html

———, "Controlled Unclassified Information," webpage, last reviewed November 13, 2017n. As of November 28, 2017:
https://www.archives.gov/cui

———, "CUI Registry—Categories and Subcategories," webpage, last reviewed November 15, 2017o. As of November 28, 2017:
https://www.archives.gov/cui/registry/category-list

National Archives and Records Administration, Controlled Unclassified Information Office, *What Is CUI? Answers to the Most Frequently Asked Questions*, Washington, D.C., 2011. As of November 28, 2017:
https://www.archives.gov/files/cui/documents/2011-what-is-cui-bifold-brochure.pdf

National Commission on Terrorist Attacks upon the United States, *The 9/11 Commission Report: Final Report of the National Commission on Terrorist Attacks upon the United States*, Washington, D.C.: U.S. Government Printing Office, July 22, 2004.

National Institute of Standards and Technology, *Guide for Mapping Types of Information and Information Systems to Security Categories*, Vol. 1, Special Publication 800-60, August 2008.

NIST—*See* National Institute of Standards and Technology.

Obama, Barack, "Freedom of Information Act: Memorandum for the Heads of Executive Departments and Agencies," memorandum for the Heads of Executive Departments and Agencies, Washington, D.C., January 26, 2009a. As of November 28, 2017:
https://www.imls.gov/sites/default/files/presidentmemorandum620.pdf

———, "Classified Information and Controlled Unclassified Information," memorandum for the Heads of Executive Departments and Agencies, Washington, D.C., May 27, 2009b. As of November 28, 2017:
https://obamawhitehouse.archives.gov/the-press-office/
presidential-memorandum-classified-information-and-controlled-unclassified-informat

OMB Watch, *Controlled Unclassified Information: Recommendations for Information Control Reform*, Washington, D.C.: Center for Effective Government, July 2009. As of October 25, 2017:
https://www.foreffectivegov.org/sites/default/files/info/2009cuirpt.pdf

Public Law 114–328, Section 901, National Defense Authorization Act for Fiscal Year 2017, December 23, 2016.

Riposo, Jessie, Megan McKernan, Jeffrey A. Drezner, Geoffrey McGovern, Daniel Tremblay, Jason Kumar, and Jerry M. Sollinger, *Issues with Access to Acquisition Data and Information in the Department of Defense: Policy and Practice*, Santa Monica, Calif.: RAND Corporation, RR-880-OSD, 2015. As of May 9, 2017:
https://www.rand.org/pubs/research_reports/RR880.html

Under Secretary of Defense, Acquisition, Technology, and Logistics, *Performance of the Defense Acquisition System: 2013 Annual Report*, Washington, D.C.: U.S. Department of Defense, June 28, 2013. As of July 11, 2017:
http://www.dtic.mil/docs/citations/ADA587235

———, *Performance of the Defense Acquisition System: 2014 Annual Report*, Washington, D.C.: U.S. Department of Defense, June 13, 2014. As of July 11, 2017:
http://www.dtic.mil/docs/citations/ADA603782

————, *Performance of the Defense Acquisition System: 2015 Annual Report*, Washington, D.C.: U.S. Department of Defense, September 16, 2015. As of July 11, 2017:
http://www.dtic.mil/docs/citations/ADA621941

————, *Performance of the Defense Acquisition System: 2016 Annual Report*, Washington, D.C.: U.S. Department of Defense, October 24, 2016. As of July 11, 2017:
http://www.dtic.mil/docs/citations/AD1019605

U.S. Computer Emergency Readiness Team, "About Us: Our Mission," webpage, undated. As of July 11, 2017:
https://www.us-cert.gov/about-us

————, *Protecting Aggregated Data*, December 5, 2005. As of November 28, 2017:
https://www.us-cert.gov/sites/default/files/publications/Data-Agg-120605.pdf

USD(AT&L)—*See* Under Secretary of Defense for Acquisition, Technology, and Logistics.

U.S. Task Force on Controlled Unclassified Information, *Report and Recommendations of the Presidential Task Force on Controlled Unclassified Information*, Washington, D.C., August 25, 2009. As of November 28, 2017:
https://www.archives.gov/files/cui/documents/2009-presidential-task-force-report-and-recommendations.pdf

Weinstein, Allen, "Establishment of the Controlled Unclassified Information Office," memorandum to the Heads of Executive Departments and Agencies, Washington, D.C., May 9, 2008.